Evolution vs. Creationism

Evolution vs. Creationism

An Introduction

Eugenie C. Scott

Foreword by Niles Eldredge

UNIVERSITY OF CALIFORNIA PRESS
BERKELEY LOS ANGELES LONDON

University of California Press
Berkeley and Los Angeles, California

University of California Press, Ltd.
London, England

First paperback edition published in 2005 by the University of California Press

Library of Congress Cataloging-in-Publication Data

Scott, Eugenie Carol, 1945–.
Evolution vs. creationism : An introduction / Eugenie C. Scott ; foreword by Niles Eldredge.
p. cm.
Originally published: Westport, Conn. : Greenwood Press, 2004.
Includes bibliographical references and index.
ISBN 0-520-24650-0 (pbk. : alk. paper)
1. Evolution (Biology). 2. Creationism. I. Title.

QH367 .S395 2005
576.8—dc22 2005048649

Manufactured in the United States of America

14 13 12 11 10 09 08 07 06
10 9 8 7 6 5 4

The paper used in this publication meets the minimum requirements of ANSI/NISO Z39.48-1992 (R 1997) (Permanence of Paper).

To my family, Charlie, Carrie, and Pat

Contents

Foreword: The Unmetabolized Darwin

A few weeks ago, I saw Darwin's name invoked in two separate articles in a single edition of *The New York Times*. One dwelt on a creationism controversy raging in a midwestern state, while the other used the expression "Darwinian" in an offhand manner to allude to the dog-eat-dog competitiveness of the business world. I found it striking that, in both instances, it was *Darwin*, and not *evolution*, that was the key word. For in the beginning of the twenty-first century, it is Charles Robert Darwin who still stands out as the towering nineteenth-century intellectual figure who still gives modern society fits. Both Sigmund Freud and Karl Marx (to choose two others whose work also shook up Western society), though far from forgotten, after a good run have begun to fade from the front pages. Darwin has recently replaced Charles Dickens on the British 10-pound note—ostensibly because his beard looks better, but in reality because he remains out front in our collective consciousness, increasingly alone among the voices of the past.

Why? Why does Darwin still bother so many of us in the Western world? Is it because Darwin's ideas of evolution are so difficult to understand? Or is it the very *idea* of evolution that is causing the problem?

The answer, of course, is the latter: the evolution of life through natural processes—and especially the recognition that our own species, *Homo sapiens*, is as inextricably linked to the rest of the living world as are redwood trees, mushrooms, sponges, and bacteria—still does not sit well with an awful lot of the citizenry of the United States and other Western countries. It is not that such skeptics are stupid—or even, at least in terms of their spokespersons, ill-informed. It's not, in other words, that creationists don't understand evolution: it's that they don't like it. Indeed, they revile it.

The reason that Darwin's name is still invoked so routinely is that social discourse on the cosmic origins of human beings has been stuck in a rut since the publication of his *On the Origin of Species* (1859). Roughly half of modern society at large grasps his point and is thereby able to understand why we look so much like chimps and orangs—similar to the way people look at the matching shorelines of South America and Africa and have no problem with the idea of continental drift. It seems commonsensical to this 50 percent of society to see us as the product of natural evolutionary processes—and when new facts come along, such as the astonishing 98.4 percent genetic similarity between humans and chimps, they seem to fit right in. These people have absorbed the evolutionary lesson and have moved on with their lives.

Darwin would be troubled but not especially surprised that the other roughly 50 percent of Americans (perhaps fewer numbers in his native England and on the European continent) still intransigently reject evolution. He had fully realized that life had evolved through natural selection—and that humans had evolved along with everything else—by the late 1830s. Yet, as is well known, Darwin pretty much kept his views a secret until virtually forced to "come out of the closet" and publish his views in the late 1850s by Alfred Russel Wallace's disclosure in a letter to Darwin that he had developed the same set of ideas. Darwin didn't want his earth-shattering idea to be scooped, so he hurriedly wrote the *Origin*—a book that sold out its initial print run on its first day of publication.

Although Darwin sometimes said that he waited 20 years to publish his ideas because he wanted to hone his concepts and marshal all the evidence he could (in itself a not-unreasonable claim), it is clear that the real reason for the delay was his fear of the firestorm of anger that his ideas were sure to unleash. His own wife was unhappy with his ideas; indeed, the marriage was almost called off when Darwin told her, against his father's advice, of his increasing religious doubts occasioned by his work. If Darwin's own faith was challenged by his conviction that life, including human life, had evolved through natural causes, he knew full well that the religiously faithful—nearly 100 percent of the population of Great Britain—would see his ideas in the very same stark terms. They too would see evolution as a challenge to the basic tenets of the Christian faith, and they would be very, very upset.

I agree with those historians who point to Darwin's nearly daily bouts with gastrointestinal upset as a manifestation of anxiety rather than of any systemic physical illness. Darwin finally did tell his new friend Joseph Hooker in 1844 a little bit about his secret ideas on evolution—telling him at the same time, though, that "it was like confessing a murder." Darwin knew he had the equivalent of the recipe for an atomic bomb, so devastating an effect would his ideas have on British society when he finally announced them. No wonder he was so hesitant to speak out; no wonder he was so anxious.

And, of course, his fears were well grounded. If it is the case that the majority of practitioners of the mainstream Judaeo-Christian religions have had little problem concluding that it is the job of science to explain the material contents of the universe and how it works, and the task of religion to explore the spiritual and moral side of human existence, it nonetheless remains as true today as it was in the nineteenth century that a literal reading of Genesis (with its two and a half non-identical accounts of the origin of the earth, life, and human beings) does not readily match up with the scientific account. There was a conflict then, and there remains a conflict today, between the scientific account of the history of earth and the evolution of life, on the one hand, and received interpretations of the same in some of the more hard-core Judaeo-Christian sects. Darwin remains unmetabolized—the very reason that his name is still so readily invoked so long after he died in 1882.

Thus it is not an intellectual issue—try as creationists will to make it seem so. Science—as many of the writings in this book make clear—cannot deal with the supernatural. Its rules of evidence require any statement about the nature of the world to be testable—to be subjected to further testing by asking the following: If this statement is true about the world, what would I expect to observe? If the predictions are borne out by experimentation or further observation, the idea is confirmed or corroborated—but never in the final analysis actually "proven." If, on the other hand, our predictions are not realized, we must conclude that our statement is in fact wrong: we have falsified it.

What predictions arise from the notion of evolution—that is, the idea that all organisms presently on earth are descended from a single common ancestor? There are two major predictions of what life should look like if evolution has happened. As Darwin first pointed out, new features appearing within a lineage would be passed along in the same or further modified form to all its descendants—but would *not* be present in other lineages that had diverged prior to the appearance of the evolutionary novelty (Darwin knew that the idea of evolution must also include the diversification of lineages, simply because there are so many different kinds of organisms on earth). Thus the prediction: more closely related organisms will share more similarities with each other than with more remotely related kin; rats and mice will be more similar to each other than they are to squirrels; but rats and mice and squirrels (united as rodents) share more similarities than any of them share with cats. In the end, there should be a single nested set of similarities linking up all of life.

This is exactly what systematic biologists and paleontologists find as they probe the patterns of similarities held among organisms—in effect testing over and over again this grand prediction of evolution. Rats, squirrels, and mice share many similarities, but with all other animals—plus fungi and many microscopic forms of life—they share a common organization of their ("eukaryotic") cells. They share even with the simplest bacteria the presence of the molecule RNA, which, along with the slightly less ubiquitous DNA, is the feature that is shared by all of life—and the feature that should be there if all life has descended from a single common ancestor.

Does this "prove" evolution? No, we don't speak of absolute proof, but we have so consistently found these predicted patterns of similarity to be there after centuries of continual research that scientists are confident that life has evolved.

The second grand prediction of the very idea of evolution is that the spectrum of simple (bacteria) to complex (multicellular plant and animal life) should be ordered through time: the earliest forms of life should be the simple bacteria; single-celled eukaryotic organisms should come next in the fossil record—and only later do the more complex forms of multicellular life arrive. That is indeed what we do find: bacteria going back at least as far as 3.5 billion years; more complex cells perhaps 2.2 billion years; and the great "explosion" of complex animal life between five and six hundred million years—a rapid diversification that nonetheless has simpler animals (e.g., sponges and cnidarians [relatives of corals and sea anemones]) preceding more complex forms (like arthropods and mollusks). Among vertebrates, fishes preceded amphibians, which in turn preceded reptiles, which came, as would be expected, before birds and mammals.

Again, evolution is not proven—but it certainly is fundamentally and overwhelmingly substantiated by the failure to falsify this prediction of increasing complexity through time.

What do creationists have to refute the very idea of evolution? They trot out a mishmash of objections to specific scientific claims; to the extent that they are testable, creationists' ideas have long been refuted. More recently, they have reverted to notions of "irreducible complexity" and "intelligent design"—ideas presented as new, but actually part of the creationist war chest before Darwin ever published the *Origin*. The fact that organisms frequently display intricate anatomies and behaviors to perform certain functions—such as flying—has inspired the claim that there must be some Intelligent Designer behind it all, that a natural process like natural selection would be inadequate to construct such exquisite complexity.

There is, of course, no scientific way to test for the existence of the Intelligent Designer; on the other hand, we can study natural selection in the wild, in the laboratory, and in mathematical simulations. We can, however, ask whether patterns of history in systems that we *know* are intelligently designed—like cars, computers, or musical instruments—resemble those of biological history. I have actually done some work along these lines—and the answer, predictably and unsurprisingly, is that the evolutionary trees of my trilobites (the fossils I study) do not resemble the trees generated by the same program for my favorite man-made objects—the musical instruments known as cornets. The reason in a nutshell is obvious: the information in biological systems is transferred almost entirely "vertically" from parent to offspring via the DNA in sperm and egg; in man-made systems, like cornets, the information is spread as much "horizontally" (as when people copy other people's ideas) as it is vertically from old master to young pupil. The details of the history of man-made objects is invariably many times more complex than what biologists find for their organisms. I think the hypothesis of intelligent design, in this sense, is indeed falsifiable—and I think we have falsified it already.

But pursuit of scientific and intellectually valid truth is not really what creationism is all about. Creationism is about maintaining particular, narrow forms of religious belief—beliefs that seem to their adherents to be threatened by the very idea of evolution. In general, it should not be anyone's business what anyone else's religious beliefs are. It is because creationism transcends religious belief and is openly and aggressively political that we need to sit up and pay attention. For in their zeal to blot evolution from the ledger books of Western civilization, creationists have tried repeatedly for well over a hundred years to have evolution either watered down, or preferably completely removed, from the curriculum of America's public schools. Creationists persistently and consistently threaten the integrity of science teaching in America—and this, of course, is of grave concern.

Perhaps someday schools in the United States will catch up to those in other developed countries and treat evolution as a normal scientific subject. Before that happens, though, people need to understand evolution, and also understand the creationism and evolution controversy. *Evolution vs. Creationism: An Introduction* is a step toward this goal, and readers will indeed learn a great deal about the scientific, religious, educational, political, and legal aspects of this controversy. Then those of us lucky enough to study evolution as a profession won't be the only ones to appreciate this fascinating field of study.

Niles Eldredge
Division of Paleontology
The American Museum of Natural History

Acknowledgments

I have always felt envious of friends who produce well-written books at regular intervals with seemingly little effort. (Michael Ruse is astonishing in this regard; I know he doesn't really publish a book every nine months, but his publication gestation is nonetheless remarkable). This book, unlike those of friends, has taken much longer than it should have, and I thank my editor at Greenwood, Kevin Downing, for his patience. I hope he will find it worth the wait! I also thank the production services of House of Equations, who tidied up some errors and missed citations that once corrected have made this book much more useful to the reader.

This book owes a great deal to the National Center for Science Education (NCSE) project staff, who have been a consistent source of good ideas and insights into the creationism/evolution controversy. Many thanks and much appreciation to Josephina Borgeson, Wesley Elsberry, Skip Evans, Alan Gishlick, Nicholas Matzke, Eric Meikle, and Jessica Moran. Former staff Abraham Kneisley, David Leitner, and Molleen Matsumura have been similarly helpful, and remain friends who continue to contribute in many ways to NCSE. Of course, none of us would get any work done if Nina Hollenberg, Phillip Spieth, and Tully Weberg weren't keeping track of the business side of NCSE. To Nina, NCSE's administrative assistant, I offer special thanks for her efficiency and organization in helping with the onerous task of tracking down permissions for the selections from the literature, braving sometimes rude secretaries and an occasionally recalcitrant Internet. Archivist Jessica Moran traced obscure citations so much more efficiently and accurately than I could have.

My indebtedness to many other students of the creation/evolution controversy will be clear upon reading the introductory chapters. I have learned much about pedagogical issues from Brian Alters, Rodger Bybee and the rest of the Biological Sciences Curriculum Study crew, and Judy Scotchmoor; about traditional creationism from John Cole, Tom McIver, the late Robert Schadewald, and William Thwaites; about the history of the controversy from Ronald Numbers, Edward Larson, and James Moore; about philosophical issues from Philip Kitcher, Michael Ruse, and Rob Pennock; and about scientific aspects of the controversy from Brent Dalrymple, Niles Eldredge, Doug Futuyma, Ken Miller, Kevin Padian, the late Art Strahler, and many others. I have acquired an appreciation for the complexity of the science and religion aspects of the controversy from many, including, to name only a few, Jack Haught, Jim Miller, and Robert John Russell.

I thank my friends in the Sandbox who have given so generously of their scholarship and advice, to say nothing of good humor, over the years. You know who you are.

I want to give an extra thank you to my colleague Alan Gishlick for assistance with illustrations, and also to NCSE member and artist Janet Dreyer for the fossil and other drawings in chapter 2, which she did beautifully at our request at the last minute. If you peruse issues of *Reports of NCSE*, you will see her whimsical and sometimes barbed covers and other artwork, which we appreciate greatly.

I thank the authors who kindly allowed me permission to reprint their essays. I have necessarily had to reduce a large number of potential topics to a smaller number treatable in a book like this, but of course there is much left unexplored. I have tried to select writings regarding these topics that honestly and clearly express the views of both antievolutionists and those who accept evolution. I especially appreciate the cooperation of authors whose views are opposed to mine, especially Henry and John Morris from the Institute for Creation Research, and Don Batten from Answers in Genesis. Phillip Johnson and Michael Denton likewise were cordial and helpful.

A very special thanks to my colleague, NCSE Deputy Director Glenn Branch, who has contributed substantially to this book from its planning to its completion. Glenn provided valuable suggestions on the organization of chapters as well as their content, and skillfully edited the whole volume, making my prose much clearer. The usefulness of this book will owe much to his efforts. Glenn also assembled most of the References for Further Exploration section, which benefited greatly from his encyclopedic appetite for books and resources and his phenomenal recall of just about everything he has ever read. He also skillfully compiled the index.

Finally, to my family, Charlie, Carrie, and Pat, who had to pick up the slack when I wasn't around to do my share, and who were extraordinarily understanding of my erratic schedule during the writing of this book.

In the end, there is no way to thank everyone to whom I am indebted for whatever useful information this book will have. Scholarship is like that. Similarly, I have no one to blame but myself for any errors, which I hope are few. With luck, the contents of this book may inspire some reader to in turn contribute to a further understanding of this vexing problem of antievolutionism, and dare we hope, contribute thereby to a solution to it.

Preface

This is not the only available book on the creationism/evolution controversy; far from it. Even if we consider only books published between 2000 and 2004, there are dozens of books promoting, discussing, or opposing creationism.

So why another book on this subject? There are books that look at this complex controversy from historical, legal, educational, political, scientific, and religious perspectives—but no single book that looks at the controversy from *all* of these views. For those needing a survey of the creationism/evolution controversy, *Evolution vs. Creationism: An Introduction* provides "one-stop shopping," so to speak, providing the reader a foundation in the ideas that have shaped this controversy. For those who are intrigued enough to pursue the topic in more detail, the additional resources will lead to a deeper understanding. Predictably, the Internet has scores of Web sites promoting various sides of the controversy, but equally predictably, sites vary enormously in their accuracy—regardless of which side the site supports. This book gives students and their teachers a foundation in science and religion that they can use as a road map to help avoid the potholes on the "information superhighway."

I have attempted to write at a level suitable to the abilities of bright high school students and college undergraduates (it's okay if others wish to read the book, too!). At the National Center for Science Education, where I work, we regularly get calls or e-mails from students (and their teachers or professors) looking for information to help in the writing of research papers on the creationism/evolution controversy; this book is a good place to begin (note to students—don't stop with just one source!). Students often flounder while attempting such assignments, lacking enough basic science (and philosophy of science) to understand why creationist critiques of evolution are resisted so strongly by scientists, and similarly lacking the theological background to understand why the claims of creationists are not uniformly accepted by religious people. The first few chapters (on science, evolution, creationism, and religion) are intended to provide the background information necessary to understand the controversy. The second section, on the history of the controversy, puts today's headlines in context; an understanding of history is essential to make sense of the current situation, which did not arise in a vacuum.

Many teachers are questioned by students about the creationism/evolution controversy in science classes, and not infrequently in nonscience classes as well. This book is intended also to be a volume to which teachers can refer students for the answers to many of their questions. It can be used for supplemental reading for classes in the sciences, or philosophy of science, and also in social studies courses (history, sociology, political science) dealing with contemporary modern problems. At the college level, this book could provide an excellent overview for the "creationism and evolution" courses now entering the curriculum at many institutions.

Unlike most other books on the creationism/evolution controversy, I have included excerpts from the creationist literature as well as rebuttals. Much of the creationist literature is not readily available except in sectarian publications and Christian bookstores, and public school libraries are properly reluctant to carry such obviously devotional literature. I have made selections from the literature that are representative of the major themes found in the creationism/evolution controversy, and have attempted to let antievolutionists speak in their own voices.

Unfortunately, proponents of Intelligent Design (ID) Creationism—Stephen Meyer, David DeWolf, Percival Davis, Dean Kenyon, Michael Behe, Jonathan Wells, Walter Bradley, Charles Thaxton, and Roger Olsen—refused, en masse, to grant me permission to reproduce their works. Through their representative at the Seattle-based ID think tank, the Discovery Institute, these authors refused permission to reprint readily available material. The major complaint was that the popular books and articles (such as opinion-editorial articles and magazine articles) that I sought to reprint would not do justice to the complexity of ID "theory." This rationale does make one wonder why such apparently inadequate works were published in the first place and continue, in several cases, to be available on or linked to from the Discovery Institute's Web site. Their "my way or the highway" response mirrors the refusal of Discovery Institute fellows to participate in the 2001 *NOVA* television series "Evolution" on PBS, on the grounds that they wanted the hour-long episode addressing the creationism/evolution controversy to focus on ID's perspective, objecting to sharing the time with proponents of Creation Science.

Consequently, most of the selections from the ID literature presented in chapters 8, 9, 10, and 12 consist of summaries of the articles I was denied permission to reprint. References to the original articles are provided, and since most of these writings are readily available on the Internet, readers can judge for themselves whether my summaries are accurate. The exception is the presence of an article by ID proponent Phillip Johnson, who cordially and promptly granted permission for me to use it. I thank him for this courtesy.

I would like to thank Henry M. Morris, John Morris, and other personnel at the Institute for Creation Research for the professional manner in which they treated my requests to use selections from ICR authors in this book. They were aware that their works would be juxtaposed with the writings of individuals who disagree with them, but they did not consider this sufficient reason to deny an honest presentation of their views. I also thank Don Batten of Answers in Genesis, who worked with me in a professional manner to resolve disagreements over selections from literature published by AIG.

There is a huge amount of creationist literature, making the question of what to include a difficult one. I organized the readings first into subject categories: physical sciences, biological sciences, legal issues, educational issues, religious issues, and topics in philosophy of science. In making selections within each of these areas, I chose readings that illustrate major ideas and themes that run through the creationism/evolution controversy.

For the physical sciences (cosmology, astronomy, and geology), I chose readings on the ubiquitous argument that the second law of thermodynamics refutes evolution, attacks on alleged weaknesses of radioisotope dating, and a historically important area of creationist thought called "Flood Geology." Positive arguments for Special Creation, as opposed to merely arguments attempting to refute evolution, are difficult to find in creationist literature; as a representative I have included a discussion of the vapor canopy model of the earth's pre-Flood atmosphere. All of the readings for these physical science themes necessarily come from the traditional Creation Science literature; Intelligent Design proponents typically do not take positions on such matters as the age of Earth or the reliability of radiometric dating. They do, however, endorse the anthropic principle, and I include readings concerning this topic.

In biological science, the major themes I have chosen to highlight are "gaps in the fossil record" (transitional fossils), the Cambrian Explosion, and the "microevolution-macroevolution" distinction. There are of course many other biological topics discussed in the creationism/evolution controversy, but these three are ubiquitous and important, and also are treated in both the traditional Creation Science and the ID literature.

In the chapter on legal issues, I illustrate the legal history of the creationism/evolution controversy by selecting readings or portions of legal cases from the period of banning evolution, of attempting to legislate equal time for evolution and Creation Science, and the current, or Neocreationist, period, in which a variety of approaches are being explored that have in common a conscious effort to circumvent the Religion Clause of the First Amendment of the Constitution.

The topics I have chosen to illustrate educational aspects of the creationism/evolution controversy include the "fairness" argument, a variant of this argument involving teaching "all the evidence," and a perennial area of controversy, the coverage of evolution in science textbooks.

The chapter on religion considers themes such as models of science and religion interaction, literalist vs. nonliteralist approaches to creation theology, and a major concern about evolution for many religious people: whether evolution necessarily compels a feeling of a loss of purpose or meaning to life.

Finally, because the creationism/evolution controversy involves science, it also involves considerations of the nature of science itself. The last chapter of selections from the literature, then, looks at the philosophy of science. What is meant by scientific "fact" as opposed to "theory"? Both traditional Creation Science and ID promote a division of science into "origins science" and "operation science." Finally, both forms of creationism are concerned with philosophical naturalism and methodological naturalism.

The juxtaposition of articles by creationists and articles by anticreationists requires a caveat, lest students be misled. Students are ill-served if in the name of "fairness" or "critical thinking" they are misled into believing that there is a controversy in the scientific world over whether evolution occurred. There is none. Although the teaching of evolution is often regarded as controversial at the K–12 level, the subject is taught matter-of-factly in every respected secular and sectarian university or college in this country, including the Baptist institution Baylor, the Mormon flagship university Brigham Young, and, of course, Catholic Notre Dame. There is scientific controversy concerning the details of *mechanisms* and *patterns* of evolution, but not over *whether* the universe has had a history measured in billions of years, nor over whether living things share a common ancestry. It would be dishonest as well as unfair to students to pretend that a public controversy over the teaching of evolution is also a scientific controversy over whether evolution occurred.

But a public controversy there is, and its complex foundation in history, science, religion, and politics will, I hope, be interesting to readers.

INTRODUCTION: The Pillars of Creationism

This book examines the creationism/evolution controversy from a broad perspective. You will read about science, religion, education, law, history, and even some current events, because all of these topics are relevant to an understanding of this controversy. In this introduction, we will examine three antievolutionist contentions that provide a framework for thinking about this complex controversy. These "Pillars of Creationism" include scientific, religious, and educational arguments, respectively, and have been central to the antievolution movement since at least the Scopes trial in 1925. As you read the following chapters and selections, it may be helpful to keep the "Pillars of Creationism" in mind.

EVOLUTION IS A "THEORY IN CRISIS"

In 1986 New Zealand physician Michael Denton wrote a book titled *Evolution: A Theory in Crisis*, which became, and remains, very popular in creationist circles. Denton claimed that there were major scientific flaws in the theory of evolution. This idea is not new: throughout the nineteenth and twentieth centuries, there was no shortage of claims that evolution scientifically was on its last legs, as documented delightfully by Glenn Morton on his Web site (http://home.entouch.net/dmd/moreandmore.htm). Of course, such claims continue to be made in the twenty-first century as well. Ironically, Denton has rejected the antievolutionary claims of some of his readers, and describes his 1986 book as opposing Darwinism (i.e., evolution through natural selection), rather than rejecting evolution itself (Denton 1999).

Through constant reiteration in creationist literature and in letters to the editor in newspapers around the country, the idea that evolution is shaky science is constantly spread to the general public, which by and large is unaware of the theoretical and evidentiary strength of evolution. Evolution as a science is discussed in chapter 2.

EVOLUTION AND RELIGION ARE INCOMPATIBLE

Darwin made two major points in *On the Origin of Species*: that living things had evolved, or descended with modification, from common ancestors, and that the mechanism of natural selection was evolution's major cause. These two components of his book often are jumbled together by antievolutionists, who argue that if natural selection can be shown to be inadequate as an evolutionary mechanism, then the idea of common descent necessarily fails. But the two constituents of Darwin's argument are conceptually and historically distinct. Common descent was accepted by both the scientific and the religious communities earlier than was the mechanism of natural selection. Further separating the two components of "Darwinism" is the fact that the religious objections to each are quite distinct. For these reasons, I will separate these two theoretical concepts in discussing religious objections to evolution.

Common Ancestry

Biblical literalists are strongly opposed to the idea of common ancestry—especially common ancestry of humans with other creatures. According to some literal interpretations of the Bible, God created living things as separate "kinds." If living things instead have descended with modification from common ancestors, the Bible would be untrue. Many biblical literalists (Young Earth Creationists, or YECs) also believe that Earth's age is measured in thousands rather than billions of years.

Yet even before Darwin published *On the Origin of Species*, there was compelling evidence for an ancient Earth and the existence of species of living things before the advent of humans. Fossils of creatures similar to but different from living forms were known, which implied that Genesis was an incomplete record of creation. More troubling was the existence of fossils of creatures not known to be alive today, raising the possibility that God allowed some of His creatures to become extinct. Did the evidence of extinction mean that God's Creation was somehow not perfect? If Earth was ancient and populated by creatures that lived before humans, death must have preceded Adam's fall—which has obvious implications for the Christian doctrine of original sin. These theological issues were addressed in a variety of ways by clergy in the nineteenth and early twentieth centuries (see chapters 3, 4, and 11, and references).

Unquestionably, evolution has consequences for traditional Christian religion. Equally unquestionably, Christian theologians and thoughtful laymen have pondered these issues and attempted to resolve the potential contradictions between traditional religion and modern science. Some of these approaches are discussed in chapter 11.

Natural Selection

"Natural selection" refers to Darwin's principal mechanism of evolution, which you will learn about in more detail in chapter 2. Those individuals in a population that (genetically) are better able to survive and reproduce in a particular environment leave more offspring, which in turn carry a higher frequency of genes promoting adaptation to that environment. Though effective in producing adaptation, natural selection is a wasteful mechanism: many individuals fall by the wayside, poorly adapted, and fail to survive and/or reproduce.

Even Christians who accept common descent may be uneasy about Darwin's mechanism of natural selection as the major engine of evolutionary change. Common ancestry itself may not be a stumbling block, but if the variety of living things we see today is primarily the result of the incredibly wasteful and painful process of natural selection, can this really be the result of actions of a benevolent God? The theodicy issue (the theological term for the problem raised by the existence of evil in a world created by a benevolent God) is a concern for both biblical literalist and nonliteralist Christians and, as discussed in chapter 6, is a major stumbling block to the acceptance of evolution by Intelligent Design Creationists (IDCs). Yet the evidence for the operation of natural selection is so overwhelming that both IDCs and YECs now accept that it is responsible for such phenomena as pesticide resistance in insects or antibiotic resistance in bacteria. YECs interpret the wastefulness of natural selection as further evidence of the deterioration of creation since the Fall of Adam. Both YECs and IDCs deny that natural selection has the ability to transform living things into different "kinds" or to produce major changes in body plans, such as the differences between a bird and a reptile.

Thus religious objections to evolution are not simple; they span a range of concerns. Religious objections to evolution are far more important in motivating antievolutionism than are scientific objections to evolution as a weak or unsupported theory.

"BALANCING" EVOLUTION ("FAIRNESS")

A third antievolution theme present as far back as the 1925 Scopes trial and continuing today is the idea that if evolution is taught, then creationism in some form should also be taught, as a matter of fairness. The "fairness" theme has, however, had many manifestations through time, largely evolving in response to court decisions (see chapters 6, 9, and 10).

"Fairness" reflects American cultural values of allowing all sides to be heard, and also a long-standing American democratic cultural tradition that assumes an individual citizen can come to a sound conclusion after hearing all the facts—and also has the right to inform elected officials of his or her opinion. Indeed, for many local and even national issues, Americans do not defer to elected and appointed officials, but vigorously debate decisions in town meetings, city council meetings, and school board meetings.

As a result, in the United States there are disputes at the local school board level over who—scientists, teachers, or members of the general public—should decide educational content. In the 1920s, the populist orator, politician, and lawyer William Jennings Bryan raged at the audacity of "experts" who would come to tell parents what to teach their children, when (as he thought) the proposed subject matter (evolution) was diametrically opposed to parental values (see chapter 4).

Many modern-day antievolutionists make this same point, arguing that conservative Christian students should not even be exposed to evolution if their religious beliefs disagree with evolution's implications. Educators and scientists argue that a student must understand evolution to be scientifically literate, and insist that the science curriculum would be deficient if evolution were omitted. Efforts to ban the teaching of evolution failed, due both to rulings by the Supreme Court and to the growth of evolution as a science (see chapters 2, 4, 5, and 9). Antievolutionists shifted their emphasis from banning evolution to having it "balanced" with the teaching of a form of creationism called "Creation Science" (see chapters 3 and 5). When this effort also failed, antievolutionists began to lobby school boards and state legislatures to "balance" evolution with the teaching of "evidence against evolution," which in content proved to be identical to Creation Science.

The perceived incompatibility of evolution with religion (especially conservative Christian theology) is the most powerful motivator of antievolutionism for individuals. However, the "fairness" concept, because of its cultural appeal, may be even more effective, for it appeals broadly across many diverse religious orientations. Even those who are not creationists may see value in being "fair" to all sides, whether or not they believe that there is scientific validity to creationist views. Scientists and teachers argue, however, that to apply "fairness" to the science classroom is a misapplication of an otherwise worthy cultural value (see chapters 9, 10, and 11).

A LOOK FORWARD

Consider these three themes, then, as you read the following chapters. Reflect on how these "Pillars of Creationism" have influenced the history of this controversy, and continue to be reflected in creationism/evolution disputes you read about in the news or see on television. Should you encounter such a local or state-level controversy, you will, I predict, easily be able to place creationist arguments into one (or more) of these categories. The following chapters will provide context for understanding these three themes as well as the creationism/evolution controversy itself.

REFERENCE CITED

Denton, Michael. 1999. Comments on Special Creationism. In *Darwinism Defeated? The Johnson-Lamoureux Debate on Biological Origins*, edited by P. E. Johnson and D. O. Lamoureux. Vancouver, BC: Regent College Publishing.

PART I

Science, Evolution, Religion, and Creationism

The creationism/evolution controversy has been of long duration in American society and shows no sign of disappearing. To understand it requires some background in the two subject areas most closely concerned with the controversy, science and religion. Within science and religion, the subareas of evolution and creationism are clearly central to the dispute.

Most people will recognize that religion and creationism are related concepts, as are science and evolution, but there also is something called "creation science," and there is even a form of religion called "scientism." In this introductory section, then, you will read about science, evolution, religion, creationism, and scientism.

These and other subjects constitute Part I of *Evolution vs. Creationism: An Introduction*. I am assuming that readers of this book will vary greatly in their understanding of these subjects, so I have tried to present material at a level that does not leave behind the beginner but has enough detail to interest a reader with more than average background in philosophy of science, evolution, or religious studies. At minimum, readers will at least know how I define and use terms which will recur throughout the book.

In the first chapter, "Science," I consider different ways of knowing, and how the way of knowing called science is especially appropriate to knowing about the natural world. Testing is the most important component in science, and different kinds of testing are discussed. In the second chapter, "Evolution," I discuss some of the basic ideas in this quite broad scientific discipline. Chapter 3, "Beliefs," discusses religion as a universal set of beliefs, with particular attention paid to origin stories and creationisms (the plural is deliberate). It also discusses naturalism as a belief. Because of the importance of the Christian religion to the creationism/evolution controversy, most of this chapter deals with Christian creationism.

CHAPTER 1

Science: "Truth Without Certainty"

We live in a universe made up of matter and energy, a *material* universe. To understand and explain this material universe is the goal of science, which is a methodology as well as a body of knowledge obtained through that methodology. As will become clear when we discuss religion, most individuals believe that the universe includes something in addition to matter and energy, but science is limited to the latter two. The methodology of science is a topic on which any college library has dozens of feet of shelves of books and journals, so obviously just one chapter won't go much beyond sketching the bare essentials of this method. Still, I will try to show how science differs from many other ways of knowing and is particularly well-suited to explaining our material universe.

WAYS OF KNOWING

Science requires the testing of explanations of the natural world against nature itself, and discarding those explanations that do not work. What distinguishes science from other ways of knowing is its reliance upon the natural world itself as the arbiter of truth. There are many things that people are interested in, are concerned about, or want to know about that science does not address. Whether the music of Madonna or Mozart is superior is of interest to many (especially parents of teenagers), but it is not something that science addresses. Aesthetics is clearly something outside of science. Similarly, literature or music might help us understand or cope with emotions and feelings in a way that science is not equipped to do. But if one wishes to know about the natural world and how it works, science is superior to other ways of knowing.

Authority

Consider for a moment some other ways of knowing about the natural world that people sometimes use. Dr. Jones says, "Male lions taking over a pride will kill young cubs." Should you believe her? You might know that Dr. Jones is a famous specialist in lion behavior who has studied lions for 20 years in the field. *Authority* leads one to believe that Dr. Jones's statement is true. In a public bathroom, I once saw a little girl of perhaps four or five years old marvel at the faucets that automatically turned on when hands were placed below the spigot. She asked her mother, "Why does the water come out, Mommy?" Her mother answered brightly, if unhelpfully, "It's magic, dear!" When we are small, we rely on the authority of our parents and other older people, but clearly authority can mislead one, as in the case of the "magic" spigots. On the other hand, if you know that Dr. Jones has witnessed several episodes of males taking over prides and killing the youngest cubs, or that in the past her statements about animal behavior have been verified, you tend to trust her authority. But authority is not something you should rely upon; Dr. Jones might be partly or even completely wrong even with lots of experience, and Mom is not always right, either (something most people figure out even before they have their own children).

Revelation

Sometimes people believe a statement because they are told it comes from a source that is unquestionable: from God, or the gods, or from some other supernatural power. Seekers of advice from the Greek oracle at Delphi believed it because the oracle supposedly received information directly from Apollo; similarly, Muslims believe the contents of the Koran were revealed to Muhammad by God; and Christians believe the New Testament is true because the authors were directly inspired by God. A problem with revealed truth, however, is that one must accept the worldview of the speaker in order to accept the statement; there is no outside referent. If you don't believe in Apollo, you're not going to trust the Delphic oracle's pronouncements; if you're not a Mormon or a Catholic, you are not likely to believe that God speaks directly to the Mormon President or the Pope. Information obtained through revelation or authority is difficult to verify because there is not an outside referent that all parties are likely to agree upon.

Logic

A way of knowing that is highly reliable is *logic*, which is the foundation for mathematics. Among other things, logic presents rules for how to tell whether something is true or false, and is extremely useful. However, logic in and of itself, with no reference to the "real world," is not complete. It is logically correct to say, "All cows are brown. Bossy is not brown. Therefore Bossy is not a cow." The problem with the statement is the truth of the premise that all cows are brown, when many are not. To know that the proposition about cows is empirically wrong even if logically true requires reference to the real world outside the logical structure of the three sentences. To say, "All wood has carbon atoms. My computer chip has no carbon atoms. Therefore my computer chip is not made of wood" is both logically and empirically true.

Science

Science does include logic—statements that are not logically true cannot be scientifically true—but what distinguishes the scientific way of knowing is the requirement of going to the outside world to verify claims. Statements about the natural world are tested against the natural world, which is the final arbiter. Of course, this approach is not perfect: one's information about the natural world comes from experiencing the natural world through the senses (touch, smell, taste, vision, hearing) and instrumental extensions of these senses (microscopes, telescopes, telemetry, chemical analysis, etc.), which can be faulty or incomplete. As a result science, more than any of the other ways of knowing described here, is more tentative in its proclamations, but ultimately this leads to more confidence in scientific understanding: the willingness to change one's explanation with more or better data, or a different way of looking at the same data, is one of the great strengths of the scientific method. The anthropologist Ashley Montagu summarized science rather nicely when he wrote, "The scientist believes in proof without certainty, the bigot in certainty without proof" (Montagu 1984: 9).

Thus science requires deciding among alternative explanations of the natural world by going to the natural world itself to test them. There are many ways of testing an explanation, but virtually all of them involve the idea of holding constant some factors that might influence the explanation, so that some alternate explanations can be eliminated. The most familiar kind of test is the *direct experiment*, which is so familiar that it is even used to sell us products on television.

DIRECT EXPERIMENTATION

Does RealClean detergent make your clothes cleaner? The smiling company representative in the TV commercial takes two identical shirts and pours something messy on each one, and drops them into identical washing machines. RealClean brand detergent goes into one machine, the recommended amount of a rival brand into the other. Each washing machine is set to the same cycle, for the same period of time, and the ad fast-forwards to show the continuously smiling salesperson taking the two shirts out. Guess which one is cleaner.

Now, it would be very easy to rig the demonstration so that RealClean does a better job: the salesperson could use less of the other detergent, use an inferior-performing washing machine, put the RealClean shirt on a soak cycle 45 minutes longer than brand X, employ different temperatures, wash the competitor's shirt on delicate rather than regular cycle—I'm sure you can think of a lot of ways that RealClean's manufacturer could ensure that its product comes out ahead. It would be a bad sales technique, however, because we're familiar with the direct experimental type of test, and someone would very quickly call "Foul!" To convince you that they have a better product, the makers of the commercial have to remove every factor that might possibly explain why the shirt came out cleaner when washed in their product. They have to hold constant or *control* all these other factors—type of machine, length of cycle, temperature of the water, and so on—so that the only reasonable explanation for the cleaner shirt is that RealClean is a better product. The experimental method—performed fairly—is a very good way to persuade people that your explanation is correct. In science, too, someone will call "Foul!" (or at least, "You blew it!") if a test doesn't consider other relevant factors.

Direct experimentation is a very powerful—as well as familiar—research design. As a result, some people think that this is the only way that science works. Actually, in science, experimentation is not as important as *testing*, and direct experimentation is only one kind of testing. The key element to testing an explanation is to hold variables constant, and one can hold variables constant in many ways other than being able to directly manipulate them (as one can water temperature in a washing machine). In fact, the more complicated the science, the less likely an experimenter is to use direct experimentation.

In some tests, variables are controlled statistically; in others, especially in biological field research or in social sciences, there are natural situations where important variables are controlled by the nature of the experimental situation itself. These observational research designs are another type of direct experimentation.

Noticing that male guppies are brightly colored and smaller than the drab females, you might wonder whether having bright colors made male guppies easier prey. How would you test this idea? If conditions allowed, you might be able to perform a *direct experiment* by moving brightly colored guppies to a high-predation environment and monitoring them over several generations to see how they do. If not, though, you could still perform an observational experiment by looking for natural populations of the same or related species of guppies in environments where predation was high and other environments where predation was low. You would also want to pick environments where the amount of food was roughly the same—can you explain why? What other environmental factors would you want to hold constant at both sites?

When you find guppy habitats that naturally vary only in the amount of predation and not in other ways, then you're ready to compare the brightness of color in the males. Does the color of male guppies differ in the two environments? If males are less brightly colored in environments with high predation, this would support the idea that brighter guppy color makes males easier prey. (What if in the two kinds of environments, male guppy color is the same?)

Indirect experimentation is used for scientific problems where the phenomena being studied—unlike the guppies—cannot be directly observed.

INDIRECT EXPERIMENTATION

In some fields, not only is it impossible to directly control the variables, but the phenomena themselves may not be directly observable. A research design known as *indirect experimentation* is often utilized in such fields. Explanations can be tested even if the phenomena being studied are too far away, too small, or too far back in time to be observed directly. For example, giant planets recently have been discovered orbiting distant stars—though we cannot directly observe them. Their presence is indicated by the gravitational effects they have on the suns around which they revolve: because of what we know about how the theory of gravitation works, we can infer that the passage of a big planet around a sun will make the sun wobble. Through the application of principles and laws in which we have confidence, it is possible to infer that these planetary giants do exist, and to make estimates of their size and speed of revolution.

Similarly, the subatomic particles studied by physicists are too small to be observed directly, but particle physicists certainly are able to test their explanations. By applying knowledge about how particles behave, they are able to create indirect experiments to test claims about the nature of particles. Let's say that a physicist wants to ascertain properties of a particle—its mass, charge, or speed. Based on observations of similar particles, he makes an informed estimate of the speed. To test the estimate, he might bombard it with another particle of known mass, because if the unknown particle has a mass of M, it will cause the known particle to ricochet at velocity V. If the known particle does ricochet as predicted, this would support the hypothesis about the mass of the unknown particle. Thus theory is built piece by piece, through inference based on accepted principles.

In truth, most scientific problems are of this "If . . . then . . ." type, whether the phenomena investigated are directly observable or not. If male guppy color is related to predation, then we should see duller males in high-predation environments. If a new drug stimulates the immune system, then individuals taking it should have fewer colds than controls. If human hunters were involved in the destruction of large Australian land mammals, we should see extinction events correlating with the appearance of the first Aborigines. We test by consequence in science all the time. Of course—because scientific problems are never solved so simply—if we get the consequence we predict, this does not mean we have "proven" our explanation. If you found that guppy color does vary in environments where predation differs, this does not mean you've proved yourself right about the relationship between color and predation. To understand why, we need to consider what we mean by proof and disproof in science.

PROOF AND DISPROOF

Proof

Scientists don't usually talk about "proving" themselves right, because "proof" suggests certainty (remember Ashley Montagu's "truth without certainty"!). The testing of explanations is in reality a lot messier than the simplistic descriptions given above. One can rarely be sure that all the possible factors that might explain why a test produced a positive result have been considered. In the guppy case, for example, let's say that you found two habitats that differed in the number of predators but were the same in terms of amount of food, water temperature, and hiding places—you tried to hold constant as many factors as you could think of. If you find that guppies are less colorful in the high-predation environment, you might think you have made the link, but some other scientist may come along and discover that your two environments differ in water turbidity. If turbidity affects predation—or the ability of female guppies to select the more colorful males—this scientist can claim that you were premature to conclude that color is associated with predation. In science we rarely claim to "prove" a theory—but positive results allow us to claim that we are likely to be on the right track. And then you or some other scientist can go out and test some more. Eventually we may achieve a consensus about guppy color being related to predation, but we wouldn't conclude this after one or a few tests. This back-and-forth testing of explanations provides a reliable understanding of nature, but the procedure is neither formulaic nor even especially tidy over the short run. Sometimes it's a matter of two steps forward, a step to the side (maybe down a blind alley), half a step back—but gradually the procedure, and with it human knowledge, lurches forward, leaving us with a clearer knowledge of the natural world and how it works.

In addition, most tests of anything other than the most trivial of scientific claims do not result in slam-dunk, now-I've-nailed-it, put-it-on-the-T-shirt conclusions, but rather in more or less tentative statements: a statement is weakly, moderately, or strongly supported, depending on the quality and completeness of the test. Scientific claims become accepted or rejected depending on how confident the scientific community is about whether the experimental results could have occurred that way just by chance—which is why statistical analysis is so important a part of most scientific tests. Animal behaviorists note that some social species share care of the offspring. Does this make a difference in the survival of the young? Some female African silver-backed jackals, for example, don't breed in a given season, but help to feed and guard the offspring of a breeding adult. If the helper phenomenon is directly related to pup survival, then more pups should survive.

One study tested this claim by comparing the reproductive success of jackal packs with and without helpers, and found that for every extra helper a mother jackal had, she successfully raised one extra pup per litter over the average survival rate (Hrdy 2001). These results might encourage you to accept the claim that helpers contribute to the survival of young, but only one test on one population is not going to be convincing. Other tests on other groups of jackals would have to be conducted to confirm the results, and to be able to generalize to other species the principle that reproductive success is improved by having a helper would require conducting tests on other social species. Such studies in fact have been performed across a large range of birds and mammals, and a consensus is emerging about the basic idea of helpers increasing survivability of the young. But there are many remaining questions, such as whether a genetic relationship always exists between the helper and either the offspring or the helped mother.

Science is quintessentially an open-ended procedure in which ideas are constantly tested, and rejected or modified. Dogma—an idea held by belief or faith—is anathema to science. A friend of mine once was asked to state how he ended up a scientist. His tongue-in-cheek answer illustrates rather nicely the nondogmatic nature of science: "As an adolescent I aspired to lasting fame, I craved factual certainty, and I thirsted for a meaningful vision of human life—so I became a scientist. This is like becoming an archbishop so you can meet girls" (Cartmill 1988: 452).

In principle, all scientific ideas may change, though in reality there are some scientific claims that are held with confidence, even if details may be modified. The physicist James Trefil (1978) suggested that scientific claims can be conceived as arranged in a series of three concentric circles (see Figure 1.1). In the center circle are the core ideas of science: the theories and facts that we have great confidence in because they work so well to explain nature. Heliocentricism, gravitation, atomic theory, and evolution would be examples. The next concentric circle outward is the frontier area of science, where research and debate are actively taking place on new theories or modifications/additions to core theories. Clearly no one is arguing with the basic principle of heliocentricism, but on the frontier, planetary astronomers still are learning things and testing ideas about the solar system. That matter is composed of atoms is not being challenged, but atomic theory is being added to and modified by the discoveries of quantum physics.

The outermost circle is the fringe, a breeding ground for ideas that very few professional scientists are spending time on: unidentified flying objects, telepathy and the like, perpetual motion machines, and so on. Generally the fringe is not a source of new ideas for the frontier, but occasionally (very occasionally!) ideas on the fringe will muster enough support to be considered by scientists and will move into the frontier. They may well be rejected and end up on the fringe again or be discarded completely, but occasionally they may move to the frontier, and perhaps eventually into the core. Continental drift began as a fringe idea, moved to the frontier as data began to accumulate in its favor, and finally became a core idea of geology when seafloor spreading was discovered to be the mechanism that pushes continents apart.

Indeed, we must be prepared to realize that even core ideas may be wrong, and that somewhere, sometime, there may be a set of circumstances that could refute even our most confidently held theory. But for practical purposes, one needn't fall into a slough of despond over the relative tentativeness of scientific explanation. That the theory of gravitation may be modified or supplemented sometime in the future is no reason to give up riding elevators (or, even less advisedly, to jump off the roof). Science gives us reliable, dependable, and workable explanations of the natural world—even if it is good philosophy of science to keep in mind that in principle anything can change.

Figure 1.1
Scientific concepts and theories can be arranged as a set of nested categories with Core ideas at the center, Frontier ideas surrounding them, and Fringe ideas at the edge (after Trefil, 1978). Courtesy of Alan Gishlick

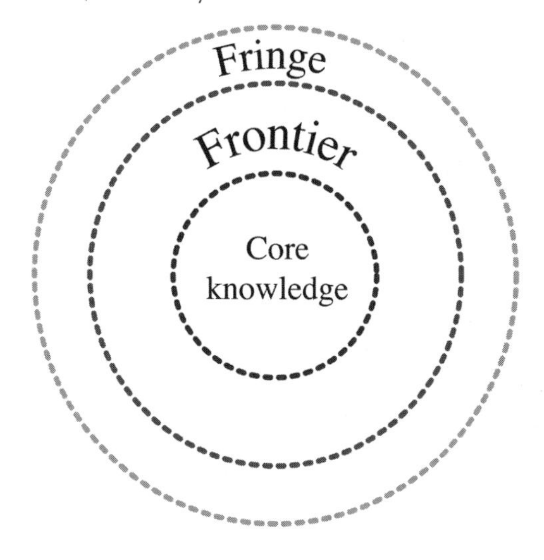

On the other hand, even if it is usually not possible absolutely to *prove* a scientific explanation correct—there might always be some set of circumstances or observations somewhere in the universe that would show your explanation wrong—to *disprove* a scientific explanation is possible. If you hypothesize that it is raining outside, and walk out the door to find the sun is shining and the ground is dry, you have indeed disproven your hypothesis (assuming you are not hallucinating). So disproving an explanation is easier than proving one true, and, in fact, progress in scientific explanation has largely come by rejecting alternate explanations. The ones that haven't been disconfirmed yet are the ones we work with—and some of those we feel very confident about.

Disproof

Now, if you are a scientist, obviously you will collect observations that support your explanation, but others are not likely to be persuaded just by a list of confirmations. Like "proving" RealClean detergent washes clothes best, it's easy to find—or concoct—circumstances that favor your view, which is why you have to bend over backward in setting up your test so that it is fair. So you set the temperature on both washing machines to be the same, you use the same volume of water, you use the recommended amount of detergent, and so forth. In the guppy case, you want to hold constant the amount of food in high-predation environments and low-predation environments, and so on. If you are wrong about the ability of RealClean to get the stains out, there won't be any difference between the two loads of clothes, because you have controlled or held constant all the other factors that might explain why one load of clothes emerged with fewer stains. You will have disproved your hypothesis about the alleged superior stain-cleaning qualities of RealClean. You are conducting a fair test of your hypothesis if you set up the test so that everything that might give your hypothesis an advantage has been excluded. If you don't, another scientist will very quickly point out your error, so it's better to do it yourself and save yourself the embarrassment!

What makes science challenging—and sometimes the most difficult part of a scientific investigation—is coming up with a testable statement. Is the African AIDS epidemic the result of tainted oral polio vaccine (OPV) administered to Congolese in the 1950s? Chimpanzees carry simian immunodeficiency virus, which researchers believe is the source for the AIDS-causing virus HIV (human immunodeficiency virus). Was a batch of OPV grown on chimp kidneys the source of African AIDS? If chimpanzee DNA could be found in the 50-year-old vaccine, that would strongly support the hypothesis. If careful analysis could not find chimpanzee DNA, that would fail to support the hypothesis, and you would have less confidence in it. Such a test was conducted, and after very careful analysis, no chimp DNA was found in samples of the old vaccine (Poinar et al. 2001).

This did not *disprove* the hypothesis that African AIDS was caused by tainted OPV (perhaps one of the other batches was tainted), but it is strong evidence against it. Again, as in most science, we are dealing with probabilities: if all four batches of OPV sent to Africa in the 1950s were prepared in the same manner, at the same time, in the same laboratory, what is the probability that one would be completely free of chimp DNA and one or more other samples tainted? Low, presumably, but because the probability is not 0 percent, we cannot say for certain that the OPV/AIDS link is out of the question. However, it has been made less plausible, and since the positive evidence for the hypothesis was thin to begin with, few people are taking the hypothesis seriously. Both disproof of hypotheses and failure to confirm are critical means by which we eliminate explanations and therefore increase our understanding of the natural world.

Now, you might notice that although I have not defined them, I already have used two scientific terms in this discussion: *theory* and *hypothesis*. You may already know what these terms mean—probably everyone has heard that evolution is "just a theory," and many times you have probably said to someone with whom you disagree, "Well, that's just a hypothesis." You might be surprised to hear that scientists don't use these terms in quite this way.

FACTS, HYPOTHESES, LAWS, AND THEORIES

How do you think scientists would rank the terms *fact*, *hypothesis*, *law*, and *theory*? How would *you* list these four from most important to least? Most people list *facts* on top, as the most important, followed by laws, then theories, with hypotheses being least important, at the bottom:

Most important
Facts
Laws
Theories
Hypotheses
Least important

You may be surprised that scientists rearrange this list, as follows:

Most important
Theories
Laws
Hypotheses
Facts
Least important

Why is there this difference? Clearly, scientists must have different definitions of these terms compared to how we use them "on the street." Let's start with facts.

Facts

If someone said to you, "List five scientific facts," you could probably do so with little difficulty. Living things are composed of cells. Gravity causes things to fall. The speed of light is about 186,000 miles/second. Continents move across the surface of Earth. Earth revolves around the sun. And so on. Scientific facts, most people think, are claims that are rock solid, about which scientists will never change their minds. Most people think that facts are just about the most important part of science, and that the job of the scientist is to collect more and more facts.

Actually, facts are useful and important, but they are far from being the most important elements of a scientific explanation. In science, facts are *confirmed observations*. After the same result is obtained after numerous observations, scientists will accept something as a fact and no longer continue to test it. If you hold up a pencil between thumb and forefinger, and then stop supporting it, it will fall to the floor. All of us have experienced unsupported objects falling; we've leaped to catch the table lamp as the toddler accidentally pulls the lamp cord. We consider it a fact that unsupported objects fall. It is always possible, however, that some circumstance may arise when a fact is shown not to be correct. If you were holding that pencil while orbiting Earth on the space shuttle and let it go, it would not fall (it would float). It also would not fall if you were on an elevator with a broken cable that was hurtling at 9.8 meters/second^2 toward the bottom of a skyscraper—but let's not dwell on that scenario. So technically, unsupported objects don't always fall, but the rule holds well enough for ordinary use. One is not frequently on either the space shuttle or a runaway elevator, or in other circumstances where the confirmed observation of unsupported items falling does not hold. It would in fact be perverse for one to reject the conclusion that unsupported objects fall just because of the existence of helium balloons!

Other scientific "facts" (confirmed observations) have been shown not to be true. For a decade or so back in the 1950s, it was thought that humans had 22 chromosome pairs, but better cell staining techniques have revealed that we actually have 23 pairs. A fact has changed, in this case with more accurate means of measurement. At one point, we had confirmed observations of 22 chromosome pairs, but now have more confirmations of 23 pairs, so we accept the latter—although at different times, both were considered "facts."

So facts are important but not immutable; they can change. An observation, though, doesn't tell you very much about how something works. It's a first step toward knowledge, but by itself it doesn't get you very far, which is why scientists put it at the bottom of the hierarchy of explanation.

Hypotheses

Hypotheses are statements of the relationship among things, often taking the form of "if . . . then . . ." statements. If brightly colored male guppies are more likely to attract predators, then in environments with high predation, guppies will be less brightly colored. If levels of lead in the bloodstream of children is inversely associated with IQ scores, then children in environments with larger amounts of lead should have lower IQ scores. Elephant groups are led by matriarchs, the eldest females. If the age (and thus experience) of the matriarch is important for the survival of the group, then groups with younger matriarchs will have higher infant mortality than those led by older ones. Each of these hypotheses is directly testable and can be either disconfirmed or confirmed (note that hypotheses are not "proved right"—any more than any scientific explanation is "proven"). Hypotheses are very important in the development of scientific explanations. Whether rejected or confirmed, tested hypotheses help to build explanations by removing incorrect approaches and encouraging the further testing of fruitful ones. Much hypothesis testing in science depends on demonstrating that a result found in a comparison occurs more or less frequently than would be the case if only chance were operating; statistics and probability are important components of scientific hypothesis testing.

Laws

There are many laws in science (e.g., the laws of thermodynamics, Mendel's laws of heredity, Newton's inverse square law, the Hardy-Weinberg law). Laws are extremely useful empirical generalizations: they state what, under certain conditions, will happen. During cell division, under Mendel's law of independent assortment, we expect genes to act like particles and separate independently of one another. Under conditions found in most places on Earth's surface, masses will attract one another in inverse proportion to the square of the distance between them. If a population of organisms is above a certain size, is not undergoing natural selection, and has random mating, the frequency of genotypes of a two-gene system will be in the proportion $p^2 + 2pq + q^2$.

Outside of science, we also use the term "law." It is the law that everyone must stop for a stoplight. Laws are uniform, and in that they apply to everyone in the society, they are universal. We don't usually think of laws changing, but of course they do: the legal system has a history, and we can see that the legal code used in the United States has evolved over several centuries from legal codes in England. Still, laws must be relatively stable or people would not be able to conduct business or know which practices will get them in trouble. One will not anticipate that if today everyone drives on the right side of the street, tomorrow everyone will begin driving on the left. Perhaps because of the stability of societal laws, we tend to think of scientific laws as also stable and unchanging.

However, scientific laws can change, or not hold under some conditions. Mendel's law of independent assortment tells us that the hereditary particles will behave independently as they are passed down from generation to generation. For example, the color of a pea flower is passed on independently from the trait for stem length. But after more study, geneticists found that Mendel's law can be "broken" if the genes are very closely associated on the same chromosome. So minimally, Mendel's law of independent assortment had to be modified in terms of new information—which is standard behavior in science. Newton's law of gravitation has to be modified (i.e., can be "broken" like Mendel's law) under conditions of a vacuum. Some laws will not hold if certain conditions are changed. Laws, then, can change just as facts can.

Laws are important, but as descriptive generalizations, they rarely *explain* natural phenomena. That is the role of the final stage in the hierarchy of explanation: *theory*. Theories *explain* laws and facts. Theories therefore are more important than laws and facts, and thus scientists place them at the top of the hierarchy of explanation.

Theories

The word "theory" is perhaps the most misunderstood word in science. In everyday usage, the synonym of theory is "guess" or "hunch." Yet according to the National Academy of Sciences, a theory is defined as "a well-substantiated explanation of some aspect of the natural world that can incorporate facts, laws, inferences, and tested hypotheses" (National Academy of Sciences 1998: 7). To explain something scientifically requires an interconnected combination of laws, tested hypotheses, and other theories. This reliance upon inferential reasoning is the hallmark of theorizing.

Many high school (and even, unfortunately, some college) textbooks describe theories as tested hypotheses, as if a hypothesis that is confirmed is somehow promoted to a theory, and a really, really good theory gets crowned as a law. Unfortunately, this is not how scientists use these terms, but most people are not scientists and scientists have not done a very good job of communicating the meanings of these terms to students and the general public. (To be honest, some scientists are not very knowledgeable about the philosophy of science! And—to be scrupulously honest—the presentation of facts, hypotheses, laws, and theories I am presenting here is very, very simplified and unnuanced, for which I apologize to philosophers of science.)

EVOLUTION AND TESTING

What about the theory of evolution? Is it scientific? Some have claimed that since no one was there to see evolution occur, studying it cannot be scientific. Indeed, no paleontologist has ever observed one species evolving into another, but as we have seen, a theory can be scientific even if its phenomena are not directly observable. Evolutionary theory is built in the same way that theory is built in particle physics or any other field that uses indirect testing. I will devote chapter 2 to discussing evolution in detail, but let me concentrate here on the question of whether it is testable—and especially if it is falsifiable.

The essence of science is the testing of explanations against the natural world. I will argue that even though we cannot observe the evolution of, say, zebras and horses from a common ancestor, evolution is indeed a science. Such hypotheses about the patterns and descent of living things can still be tested.

The "big idea" of biological evolution (as will be discussed more fully in the next chapter) is "descent with modification." Evolution is a statement about history and refers to something that happened, to the branching of species through time from common ancestors. The pattern that this branching takes and the mechanisms that bring it about are other components of evolution. We can therefore look at the testing of evolution in three senses: Can the "big idea" of evolution (descent with modification, common ancestry) be tested? Can the pattern of evolution be tested? Can the mechanisms of evolution be tested?

Testing the Big Idea

Hypotheses about evolutionary phenomena are tested just like hypotheses about other scientific topics: the trick (as in most science!) is to figure out how to formulate your question so it can be tested. The big idea of evolution—that living things have shared common ancestors—can be tested using the "if . . . then . . ." approach—testing by consequences—used by all scientists. The biologist John A. Moore suggested a number of these "if . . . then . . ." statements that could be used to test whether evolution occurred:

1. If living things descended with modification from common ancestors, then we would expect that "species that lived in the remote past must be different from the species alive today" (Moore 1984: 486). When we look at the geological record, this is indeed what we see. There are a few standout species that seem to have changed very little over hundreds of millions of years, but the rule is that the farther back in time one looks, the more creatures differ from present forms.

2. If evolution occurred, we "would expect to find only the simplest organisms in the very oldest fossiliferous [fossil-containing] strata and the more complex ones to appear in more recent strata" (Moore 1984: 486). Again going to the fossil record, we find this is true. In the oldest strata, we find single-celled organisms, then simple multicelled organisms, and then simple versions of more complex invertebrate multicelled organisms (during the early Cambrian period). In later strata, we see the invasion of the land by simple plants, and then the evolution of complex seed-bearing plants, and then the development of the land vertebrates.

3. If evolution occurred, then "there should have been connecting forms between the major groups (phyla, classes, orders)" (Moore 1984: 489). To test this requires going again to the fossil record, but matters are complicated by the fact that not all connecting forms have the same probability of being preserved. For example, connecting forms between the very earliest invertebrate groups (which all are marine) are less likely to be found because of their soft bodies, which do not preserve as well as hard body parts such as shells and bones that can be fossilized. These early invertebrates also lived in shallow marine environments, where the probability of a creature's preservation is different depending on whether it lived under or on the surface of the seafloor: surface-living forms have a better record of fossilization due to surface sediments being glued together by bacteria. Fossilized burrowing forms haven't been found—although their burrows have. Connections between vertebrate groups might be expected to be found, because vertebrates are large animals with large calcium-rich bones and teeth that have a higher probability of fossilization than the soft body parts of the earliest invertebrates. There are, in fact, good transitions between fish and amphibians, and there are especially good transitions between reptiles and mammals. More and more fossils are being found that show structural transitions between reptiles (dinosaurs) and birds. Within a vertebrate lineage, there are often fossils showing good transitional structures. We have good evidence of transitional structures showing the evolution of whales from land animals, and modern, large, single-hoofed horses from small, three-toed ancestors. Other examples can be found in reference books on vertebrate evolution such as Carroll (1998) or Prothero (1998).

In addition to the "if . . . then . . ." statements predicting what one would find if evolution occurred, one can also make predictions about what one would *not* find. If evolution occurred and living things have branched off the tree of life as lineages split from common ancestors, one would *not* find a major branch of the tree totally out of place. That is, if evolution occurred, paleontologists would not find mammals in the Devonian age of fishes, or seed-bearing plants back in the Cambrian. Geologists are daily examining strata around the world as they search for minerals, or oil, or other resources, and at no time has a major branch of the tree of life been found seriously out of place. Reports of "man tracks" being found with dinosaur footprints have been shown to be carvings, or eroded dinosaur tracks, or natural erosional features. If indeed there had not been an evolutionary, gradual emergence of branches of the tree of life, then there is no scientific reason why all strata would not show remains of living things all jumbled together.

In fact, one of the strongest sources of evidence for evolution is the consistency of the fossil record around the world. Similarly, the fact that when we look at the relationships among living things, we see that it is possible to group organisms in gradually broader classifications. There is a naturally occurring hierarchy of organisms that has been recognized since the seventeenth century: species can be grouped into genera, genera can be grouped into families, and on into higher categories. The splitting process of evolution generates hierarchy; the fact that animals and plants can be arranged in a "tree of life" is predicted by and explained by the inference of common descent.

Not only the "big idea" of evolution can be tested; so can more specific claims within that big idea. Such claims concern pattern and process, which require explanations of their own.

Pattern and Process

Pattern. Consider that if evolution is fundamentally an aspect of history, then certain things happened and other things didn't. It is the job of evolutionary biologists and geologists to reconstruct the past as best they can, and try to ascertain what actually happened as the tree of life developed and branched. This is the *pattern* of evolution, and indeed, along with the general agreement about the gradual appearance of modern forms over the last 3.8 billion years, the scientific literature is replete with disputes among scientists about specific details of the tree of life, about which structures represent transitions between groups and how different groups are related. For instance, whales are known to be related to the group of hoofed mammals called artiodactyls, but are they more closely related to the hippopotamus branch of artiodactyls (suggested by molecular data) or the cattle branch (suggested by skeletal data)? Morphologically, most Neanderthal physical traits can be placed within the range of variation of living humans, but there are tests on fossil mitochondrial DNA that suggest modern humans and Neanderthals shared a common ancestor very, very long ago—no more recently than 300,000 years ago (Ovchinnikov et al. 2000). So are Neanderthals ancestral to modern humans, or not? There is plenty of room for argument about exactly what happened in evolution! But how do you test such statements?

Tests of hypotheses of relationship commonly use the fossil record, and it is the source of most of our conclusions about the relationships within a group. Unfortunately, sometimes one has to wait a long time before hypotheses can be tested. The fossil evidence has to exist (i.e., be capable of being preserved and actually *be* preserved), be discovered, and then painstakingly (and expensively) extracted. Only then can the analysis begin. Fortunately, we can test hypotheses about evolution—and the idea of descent with modification itself—using not only the fossil record but also anatomical, embryological, or biochemical evidence from living groups. One reason why evolution—the inference of common descent—is such a robust scientific idea is that so many different sources of information lead to the same conclusions.

We can use different sources of information to test a hypothesis about the evolution of the first primitive amphibians that colonized land. There are two main types of bony fish: the very large group of familiar ray-finned fish (fish such as trout, salmon, and sunfish) and the lobe-finned fish, represented today by only three species of lungfish and one species of coelacanth. In the Devonian, though, there were 19 families of lungfish and 3 families of coelacanths. Because of their many anatomical specializations, we know that ray-finned fish are not part of tetrapod (four-legged land vertebrate) ancestry; we and all other land vertebrates are descended from the lobefin line. Early tetrapods and lobefins both had teeth with wrinkly enamel, and shared characteristics of the shoulder girdle and jaws, plus a sac off the gut used for breathing (Prothero 1998: 358). But are we tetrapods more closely related to lungfish or coelacanths? Is the relationship among these three groups more like Figures 1.2A or 1.2B? We can treat the two diagrams as hypotheses and examine data from comparative anatomy, the fossil record, biochemistry, and embryology to confirm or discomfirm A or B.

Anatomical and fossil data support hypothesis B (Thomson 1994). Studies on the embryological development of tetrapod and fish limbs also support hypothesis B. Now, when contemplating Figure 1.2, remember that these two diagrams omit the many known fossil forms and show only living groups. It isn't that tetrapods evolved from

Figure 1.2
Are tetrapods more closely related to lungfish or to coelacanths? Courtesy of Alan Gishlick

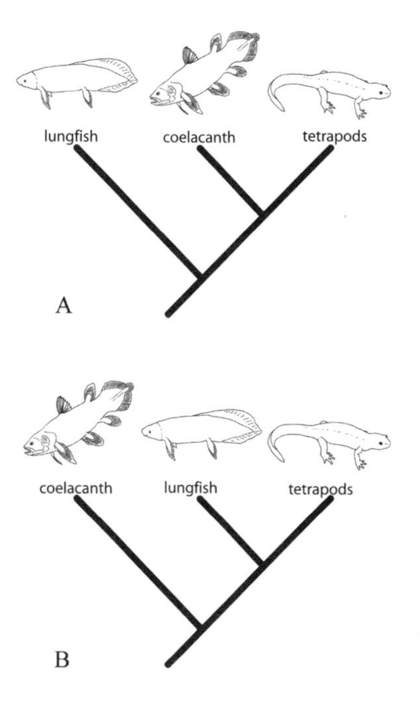

lungfish, of course, but that lungfish and tetrapods shared a common ancestor, and shared a common ancestor with one another more recently than they shared a common ancestor with coelacanths. There is a large series of fossils filling the morphological gaps between ancestors of lungfish and tetrapods (Carroll 1998).

Another interesting puzzle about the pattern of evolution is ascertaining the relationships among the phyla, which are very large groupings of kinds of animals. All the many kinds of fish, amphibians, reptiles, birds, and mammals are lumped together in one phylum (Chordata) with some invertebrate animals such as sea squirts and the wormlike lancelet (amphioxus). Another phylum (Arthropoda) consists of a very diverse group of invertebrates that includes insects, crustaceans, spiders, millipedes, horseshoe crabs, and the extinct trilobites. So you can see that phyla contain a lot of diversity. Figuring out how such large groups might be related to one another is a challenging undertaking.

Phyla are diagnosed based on basic anatomical body plans—the presence of such features as segmentation, possession of shells, possession of jointed appendages, and so forth. Fossil evidence for most of these transitions is not presently available, so scientists have looked for other ways to ascertain relationships among these large groups. The recent explosions of knowledge in molecular biology and of developmental biology are opening up new avenues to test hypotheses of relationships—including those generated from anatomical and fossil data. Chordates for a long time have been related to echinoderms based on anatomical comparisons (larvae of some echinoderms are very similar to primitive chordates) and now are further linked through biochemical comparisons (ribosomal RNA) (Runnegar 1992). Ideas about the pattern of evolution can be and are being tested.

Process. Scientists studying evolution want to know not only the pattern of evolution but also the processes behind it: the mechanisms that cause cumulative biological change through time. The most important is natural selection (discussed in chapter 2), but there are other mechanisms (mostly operating in small populations, like genetic drift) that also are thought to bring about change. One interesting current debate, for example, is over the role of genetic factors operating early in embryological development. How important are they in determining differences among—and the evolution of—the basic body plans of living things? Are the similarities of early-acting developmental genes in annelid worms and in primitive chordates like amphioxus indicative of a shared common ancestry? Another debate has to do with the rate and pace of evolution: Do changes in most lineages proceed slowly and gradually, or do most lineages remain much the same for long periods that once in a while are "punctuated" with periods of rapid evolution? We know that individuals in a population compete with each other, and that populations of a species may outbreed each other, but can there be natural selection between lineages of species through time? Are there rules that govern the branching of a lineage through time? Members of many vertebrate lineages have tended to increase in size through time; is there a general rule governing size or other trends? All of these issues and many more constitute the process of evolution. Researchers are attempting to understand these processes by testing statements against the fossil and geological records as well as other sources of information.

Natural selection and other genetically based mechanisms are regularly tested and regularly prove robust. By now there are copious examples of natural selection operating in our modern world, and it is not unreasonable to extend their operation into the past. Farmers and agricultural experts are very aware of natural selection as insects, fungi, and other crop pests become resistant to chemical controls. Physicians similarly are very aware of natural selection as they try to counter antibiotic-resistant microbes. The operation of natural selection is not disputed in the creation/evolution controversy: both supporters and detractors of evolution accept that natural selection works. Creationists, however, claim that natural selection cannot bring about differences from one "kind" to another.

Pattern and process are both of interest in evolutionary biology, and each can be evaluated independently. Disputes about the pattern of evolutionary change are largely independent of disputes about the process. That is, arguments among specialists about how fast evolution can operate, or whether it is gradual or punctuated, are irrelevant to arguments over whether Neanderthals are ancestral to modern Europeans and vice versa. Similarly, arguments about either process or pattern are irrelevant to *whether* evolution took place (the "big idea" of descent with modification). This is relevant to the creation/evolution controversy because some of the arguments about pattern or process are erroneously used to support the claim that descent with modification did not occur. Such arguments confuse different levels of understanding.

CREATIONISM AND TESTING

The topic of religion constitutes chapter 3, and creationism is a religious concept. Religion will be defined as a set of ideas concerning a nonmaterial reality; thus it would appear that—given science's concern for material explanations—science and creationism would have little in common. Yet the controversy that this book considers, the creationism/evolution controversy, includes the claim made by some that creationism is scientific, or can be made scientific, or has scientific elements. The question naturally arises, then, "Is creationism testable?"

As discussed, science operates by testing explanations of natural phenomena against the natural world. Explanations that are disproved are rejected; explanations that are not disproved—that are corroborated—are provisionally accepted (though at a later time they may be rejected or modified with new information). An important element of testing is being able to hold constant some of the conditions of the test, so that a causative effect can be correctly assigned.

The ultimate statement of creationism—that the present universe came about as the result of the action or actions of a divine Creator—is thus outside the abilities of science to test. If there is an omnipotent force in the universe, it would by definition be impossible to hold constant (to control) its effects. A scientist could control for the effects of temperature, light, humidity, or predators—but it would be impossible to control the actions of God!

The question of whether God created cannot be evaluated by science. Most believers conceive of God as omnipotent, so He can create everything all at once, a theological position known as "special creationism"; or He can create through the process of natural law, a theological position known as "theistic evolution." An omnipotent being could create the universe to appear as if it had evolved, but actually have created everything five minutes ago. The reason that the ultimate statement of creationism cannot be tested is simple: any action of an omnipotent Creator is compatible with any and all scientific explanations of the natural world. The methods of science cannot choose among the possible actions of an omnipotent Creator.

Science is thus powerless to test the ultimate claim of creationism, and must be agnostic about whether God did or did not create the material world. However, some types of creationism go beyond the basic statement "God created" to make claims of fact about the natural world. Many times these fact claims, such as those concerning the age of Earth, are greatly at variance with observations of science, and creationists sometimes evoke scientific support to support these fact claims. One creationist claim, for example, is that the Grand Canyon was laid down by the receding waters of Noah's Flood. In cases like this, scientific methods *can* be used to test creationist claims, because the claims are claims of fact. Of course, it is always possible to claim that the Creator performed miracles (that the Grand Canyon stratigraphy—which virtually all geologists consider to be impossible to have been laid down during a year's time—was created through the special actions of an omnipotent Creator), but at this point one passes from science to some other way of knowing. If fact claims are made—assuming the claimer argues scientific support for such claims—then such claims can be tested by the methods of science; some scientific views are better supported than others, and some will be rejected as a result of comparing data and methodology. But such occasions leave the realm of science for that of religion if miracles are invoked.

CONCLUSION/SUMMARY

Science is an especially good way of knowing about the natural world. It involves testing explanations against the natural world, discarding the ones that don't work and provisionally accepting the ones that do.

Theory-building is the goal of science. Theories explain natural phenomena and are logically constructed of facts, laws, and confirmed hypotheses. Knowledge in science, whether expressed in theories, laws, tested hypotheses, or facts, is provisional, though reliable. Although any scientific explanation may be modified, there are core ideas of science that have been tested so many times that we feel very confident about them and believe that there is an extremely low probability of their being discarded. The willingness of scientists to modify their explanations (theories) is one of the strengths of the method of science, and it is the major reason that knowledge of the natural world has increased exponentially over the last couple of hundred years.

Evolution, like other sciences, requires that natural explanations be tested against the natural world. Indirect observation and experimentation, involving "if . . . then . . ." structuring of questions and testing by consequence are the normal mode of testing in sciences such as particle physics and evolution, where phenomena cannot be directly observed.

The three elements of biological evolution—descent with modification, the pattern of evolution, and the process or mechanisms of evolution—can all be tested through the methods of science. The heart of creationism—that an omnipotent being created—is not testable by science, but fact claims about the natural world made by creationists can be.

In the next chapter, I will turn to the science of evolution itself.

REFERENCES CITED

Carroll, Robert L. 1998. *Vertebrate Paleontology and Evolution*. New York: W. H. Freeman.

Cartmill, Matt. 1988. Seventy-five Reasons to Become a Scientist: *American Scientist* Celebrates Its Seventy-fifth Anniversary. *American Scientist* 76: 450–463.

Hrdy, Sarah Blaffer. 2001. Mothers and Others. *Natural History*, May: 50–62.

Montagu, M. F. Ashley. 1984. *Science and Creationism*. New York: Oxford University Press.

Moore, John A. 1984. Science as a Way of Knowing—Evolutionary Biology. *American Zoologist* 24 (2): 467–534.

National Academy of Sciences. 1998. *Teaching About Evolution and the Nature of Science*. Washington, DC: National Academy Press.

Ovchinnikov, I. V., A. Gotherstrom, G. P. Romanova, V. M. Kharitonov, K. Liden, and W. Goodwin. 2000. Molecular Analysis of Neanderthal DNA from the Northern Caucasus. *Nature* 404: 490–493.

Poinar, Hendrik, Melanie Kuch, and Svante Pääbo. 2001. Molecular Analysis of Oral Polio Vaccine Samples. *Science* 292 (5517): 743–744.

Prothero, Donald R. 1998. *Bringing Fossils to Life: An Introduction to Paleontology*. Boston: WCB McGraw-Hill.

Runnegar, Bruce. 1992. Evolution of the Earliest Animals. In *Major Events in the History of Life*, edited by J. W. Schopf. Boston: Jones and Bartlett.

Thomson, Keith Stewart. 1994. The Origin of the Tetrapods. In *Major Features of Vertebrate Evolution*, edited by D. R. Prothero and R. M. Schoch. Pittsburgh, PA: The Paleontological Society.

Trefil, James. 1978. A Consumer's Guide to Pseudoscience. *Saturday Review*, April 29: 16–21.

CHAPTER 2

Evolution

EVOLUTION BROAD AND NARROW

It has been my experience as both a college professor and a longtime observer of the creationism/evolution controversy that most people have a definition of evolution rather different from that of scientists. To the question "What does evolution mean?" most people will answer, "Man evolved from monkeys" or "molecules to man." Setting aside the sex-specific language (surely no one believes that only males evolved; reproduction is challenging enough without trying to do it using only one sex), both definitions are much too narrow. Evolution involves far more than just human beings and, for that matter, far more than just living things.

The broad definition of evolution is "a cumulative change through time," and refers to the fact that the universe has had a history—that if we were able to go back into time, we would find different stars, galaxies, planets, and different forms of life on Earth. Stars, galaxies, planets, and living things have changed through time. There is astronomical evolution, geological evolution, and biological evolution. Evolution, far from being "Man evolved from monkeys," is thus integral to astronomy, geology, and biology. As we will see, it is relevant to physics and chemistry as well.

Evolution needs to be defined more narrowly within each of the scientific disciplines because both the phenomena studied and the processes and mechanisms of cosmological, geological, and biological evolution are different. Astronomical evolution deals with cosmology: the origin of elements, stars, galaxies, and planets. Geological evolution is concerned with the evolution of our own planet: its origin and its cumulative changes through time. Mechanisms of astronomical and geological evolution involve the laws and principles of physics and chemistry: thermodynamics, heat, cold, expansion, contraction, erosion, sedimentation, and the like. In biology, evolution is the inference that living things share common ancestors and have, in Darwin's words, "descended with modification" from these ancestors. The main—but not the only—mechanism of biological evolution is natural selection. Although biological evolution is the most contentious aspect of the teaching of evolution in public schools, some creationists raise objections to astronomical and geological evolution as well.

ASTRONOMICAL AND CHEMICAL EVOLUTION

Cosmologists conclude that the universe as we know it today originated from the "big bang" explosion that erupted from a compact, extremely dense mass and spread outward in all directions. Astronomers have found evidence that stars evolved from gravitational effects on swirling gases left over from the big bang. Stars combined into galaxies, the total number of which is only dimly perceived (Silk 1994). Beginning with helium, heavier elements were formed in the energy-rich cores of young stars through atomic fusion. Thus the elements were formed over approximately 10–12 billion years.

Cosmologists and geologists tell us that between 4 and 5 billion years ago the planet Earth formed from the accumulation of matter encircling our sun. In earlier times Earth looked far different from what we see today: an inhospitable place scorched by radiation, bombarded by meteorites and comets, and belching noxious chemicals from volcanoes and massive cracks in the planet's crust. Yet it is hypothesized that Earth's atmosphere evolved from that outgassing, and water might well have been brought to our planet's surface by those comets crashing into it.

Earth was bombarded by meteors and comets until about 3.8 billion years ago. In such an environment, life could not have survived. Shortly after the bombardment ceased, however, primitive replicating structures appeared. Currently, there is not yet a consensus about how these first living things originated, and there are several directions of active research. Before there were living creatures, of course, there had to be organic (carbon-containing) molecules. Presently, these molecules are produced only by living things, so the question of chemical "prebiotic" evolution involves developing plausible scenarios for the emergence of organic molecules such as sugars, purines, and pyrimidines, and the building blocks of life, amino acids.

To explore this question, in the 1950s scientists began experimenting to see whether organic compounds could be formed from methane, ammonia, water vapor, and hydrogen gases that researchers believed would have been present in Earth's early atmosphere. By electrically sparking combinations of gases, they were able to produce most of the amino acids that occur in proteins—and also the same amino acids that are found in meteorites—as well as other organic molecules (Miller 1992: 19). Because the actual composition of Earth's early atmosphere is not known, investigators have tried sparking various combinations of gases as well as the original blend. These also produce amino acids (Rode 1999: 774). Apparently, organic molecules form spontaneously on Earth and elsewhere, leading one investigator to conclude, "There appears to be a universal organic chemistry, one that is manifest in interstellar space, occurs in the atmospheres of the major planets of the solar system, and must also have occurred in the reducing atmosphere of the primitive Earth" (Miller 1992: 20).

For life to emerge, some organic molecules had to be formed and then combined into amino acids and proteins, while other organic molecules had to be combined into something that could replicate: some material that could pass information from generation to generation. Modern living things are composed of cells, and cells are set off from their environments and are recognizable entities because their constituent parts are surrounded by membranes. Much origin-of-life research focuses on explaining the origin of proteins, heredity material, and membranes.

Origin-of-life researchers joke about their models falling into two camps: "heaven" and "hell." "Hellish" theories for the origin of life point to the present-day existence of some of the simplest known forms of life in severe environments, both hot and cold. Some primitive forms of life live in hot deep-sea vents where sulfur compounds and heat provide the energy used to carry on metabolism and reproduction. Could such an environment have been the breeding ground of the first primitive forms of life? Other scientists are discovering that primitive bacteria can be found in permanently or nearly permanently frozen environments in the Arctic and Antarctic. Perhaps deep in ice or deep in the sea, protected from harmful ultraviolet radiation, organic molecules assembled into primitive replicating structures.

More "heavenly" theories note that organic molecules occur spontaneously in dust clouds of space and that amino acids have been found in meteorites. Perhaps these basic components of life were brought by these rocky visitors from outer space and combined in Earth's waters to form replicating structures.

ORIGIN OF LIFE

Whether the proponents of hell or heaven finally convince their rivals as to the most plausible scenario for the origin of the first replicating structures, it is clear that the origin of life is not a simple issue. One problem is the definition of life itself. From the ancient Greeks up through the early nineteenth century, people from European cultures believed that living things possessed an élan vital or vital spirit—a quality that sets them apart from dead things and nonliving things such as minerals or water. Organic molecules, in fact, were thought to differ from other molecules because of the presence of this spirit. This view was gradually abandoned in science when more detailed study on the structure and functioning of living things repeatedly failed to discover any evidence for such an élan vital, and when it was realized that organic molecules could be synthesized from inorganic chemicals. Vitalistic ways of thinking persist in some East Asian philosophies, such as in the concept of *chi*.

So how do we define life, then? According to generally agreed-upon scientific definition, if something is "living," it is able to acquire and use energy and to reproduce. The simplest living things today are primitive bacteria, enclosed by a membrane and not containing very many "moving parts." But they can take in and use energy, and they can reproduce by division. Even this definition is fuzzy, though: What about viruses? The microscopic viruses are hardly more than hereditary material in a packet—a protein shell. Are they alive? Well, they reproduce. They sort of use energy, in the sense that they take over a cell's machinery to duplicate their own hereditary material. But they can also form crystals, which no living thing can do, so biologists are divided over whether viruses are "living" or not. They tend to be treated as a separate, special category.

If life itself is difficult to define, you can see why explaining its origin is also going to be difficult. The first cell would have been more primitive than the most primitive bacterium known today, which itself is the end result of a long series of events: no scientist thinks that it popped into being with all its components present and functioning! A simple bacterium is "alive": it takes in energy that enables it to function, and it reproduces (in particular, it duplicates itself through division). We can recognize that a bacterium can do these things because the components that process the energy and that allow it to divide are enclosed within a membrane; we can recognize a bacterium as an entity: a cell that has several components that in a sense "cooperate." What if there were a single structure that was *not* enclosed by a membrane, but that nonetheless could conduct a primitive metabolism? Would we consider it "alive"? It is beginning very much to look like the origin of life was not a sudden event, but a continuum of events producing structures that early in the sequence we would agree are not "alive," and at the end of the sequence we would agree *are* alive, with a lot of iffy stuff in the middle.

We know that virtually all life on Earth today is based on DNA, a chainlike molecule that directs the construction of proteins and enzymes, which in turn produce creatures composed of one cell or trillions. DNA instructs cellular structures to link amino acids in a particular order to form a particular protein or enzyme. It also is the material of heredity, being passed from generation to generation. The structure of DNA is rather simple, considering all it does. DNA codes for amino acids use a "language" of four letters—A (adenine), T (thymine), C (cytosine), and G (guanine)—which, combined three at a time, determine the amino acid order of a particular protein. For example, CCA codes for the amino acid proline, and AGU for the amino acid serine. The exception to the generalization that all life is based on DNA is viruses, which can be composed of strands of RNA, another chainlike molecule that is quite similar to DNA. Like DNA, RNA is based on A, C, and G, but uses uracil (U) rather than thymine.

The origin of DNA and proteins is thus of considerable interest to origin-of-life researchers. How did the components of RNA and DNA assemble into these structures? One theory is that clay or calcium carbonate—both latticelike structures—could have provided a foundation upon which primitive chainlike molecules could have formed (Hazen et al. 2001). Because RNA is one-stranded rather than two-stranded like DNA, some scientists are building theory around the possibility of a simpler RNA-based organic "world" preceding our current DNA "world" (Joyce 1991; Lewis 1997), and very recently there has been speculation that an even simpler but related chainlike molecule, PNA (peptide nucleic acid) may have preceded the evolution of RNA (Nelson et al. 2000). Where did RNA or PNA come from? In a series of experiments combining chemicals available on the early Earth, scientists have been able to synthesize purines and pyrimidines, which form the backbones of DNA and RNA (Miller 1992).

After a replicating structure evolved (whether it was based on PNA or RNA or DNA), the structure had to be enclosed in a membrane and to acquire other bits of machinery to process energy and perform other tasks. The origin of life is a complex but active research area with many interesting avenues being investigated, though there is not yet consensus on the sequence of events that led to living things. But at some point in Earth's early history, perhaps as early as 3.8 billion years ago, life in the form of simple single-celled organisms appeared. Once life evolved, biological evolution became possible.

Although some people confuse the origin of life itself with evolution, the two are conceptually separate. Biological evolution is defined as the descent of living things from ancestors from which they differ. Life had to precede evolution! Regardless of how the first replicating molecule appeared, we see in the subsequent historical record the gradual appearance of more complex living things, and many variations on the many themes of life. We know much more about evolution than about the origin of life.

BIOLOGICAL EVOLUTION

Biological evolution is a subset of the general idea that the universe has changed through time. In the nineteenth century, Charles Darwin spoke of "descent with modification," and that phrase still nicely communicates the essence of biological evolution. "Descent" connotes heredity, and indeed, members of species pass genes from generation to generation. "Modification" connotes change, and indeed, the composition of species may change through time. Descent with modification refers to a genealogical relationship of species through time. Just as an individual genealogy theoretically can be traced back through time, so too can the genealogy of a species. And just as an individual genealogy has "missing links"—ancestors whose names or other details are uncertain—so too the history of a species is understandably incomplete. Evolutionary biologists are concerned both with the history of life—the tracing of life's genealogy—and the processes and mechanisms that produced the tree of life. This distinction between the patterns of evolution and the processes of evolution is relevant to the evaluation of some of the criticisms of evolution that will emerge later in this book. First, let's look briefly at the history of life.

The History of Life

Deep Time. The story of life unfurls against a backdrop of time—deep time: the length of time the universe has existed, the length of time that Earth has been a planet, the length of time that life has been on Earth. We are better at understanding things that we can have some experience of, but we have, and can have, no experience of deep time. Most of us can relate to a period of 100 years; a person in his fifties might reflect that 100 years ago, his grandmother was a young woman. A person in her twenties might be able to imagine what life was like for a great-grandparent 100 years ago. Thinking back to the time of Jesus, 2,000 years ago, is more difficult; although we have written descriptions of people's houses, clothes, and how they made their living, there is much we don't know of official as well as everyday life. The ancient Egyptians were building pyramids 5,000 years ago, and their lives and way of life are known in only the sketchiest outlines.

And yet the biological world of 5,000 years ago was virtually identical to ours today. The geological world 5,000 years ago would be quite recognizable: the continents would be in the same places, the Appalachians and Rocky Mountains would look pretty much as they do today, and major features of coastlines would be identifiable. Except for some minor remodeling of Earth's surface due to volcanoes and earthquakes, the filling in of some deltas due to the deposition of sediments by rivers, and some other small changes, little has changed, geologically. But our planet and life on it are far, far older than 5,000 years. We need to measure the age of Earth and the time spans important to the history of life in *billions* of years, a number that we can grasp only in the abstract.

A second is a short period of time. Sixty seconds make up a minute, and 60 minutes make up an hour. There are therefore 3,600 seconds in an hour, 86,400 in a day, 604,800 in a week, and 31,536,000 in a year. But to count to a *billion* seconds at the rate of one a second, you would have to count night and day for approximately 31 years and eight months. The age of Earth is 4.5 billion *years*, not seconds. It is an enormous amount of time. As Stephen Jay Gould remarked, "An abstract, intellectual understanding of deep time comes easily enough—I know how many zeros to place after the 10 when I mean billions. Getting it into the gut is quite another matter. Deep time is so alien that we can really only comprehend it as metaphor" (Gould 1987: 3).

Figure 2.1 presents divisions of geological time used to understand geological and biological evolution. The solar system formed approximately 4.6 billion years ago; Earth formed around 4.5 billion years ago, and life may have begun appearing relatively soon after that. During our planet's early years, however, it was bombarded regularly by comets and meteorites, as was the moon. This may have impeded the development of life: the first evidence of bacteria-like fossils dates from around 3.5 billion years ago, about half a billion years after the bombardment of Earth stopped. As discussed in the section "Astronomical and Chemical Evolution," there was a period of about half a billion years of chemical evolution before the first structures that one might consider "alive" appeared on Earth: primitive bacteria.

After these first living things appeared between 3.5 and 4 billion years ago, life continued to remain simple for more than 2 billion years. Single-celled living things bumped around in water, absorbed energy, and divided—if they weren't absorbed by some other organism first. Reproduction was asexual: when a cell divided, the result was almost always two identical cells. Very slow changes occur with asexual reproduction, and this is probably an important reason that the evolution of life moved so slowly during life's first few billion years.

Nucleated Cells. The first cells lacked a nucleus. Nucleated (eukaryotic) cells didn't evolve until about 1.5 billion years ago. In eukaryotes, the nucleus contains the DNA, which is important to cell division and to directing the processes that living things need to carry out to survive. Around 2 billion years ago, great changes in Earth's surface were taking place, including movement of continents, increases in the amount of oxygen in the atmosphere, and the establishment of an ozone shield that protected living things against ultraviolet radiation (Strickberger 1996: 168). The increase and spread of photosynthesizing single-celled organisms is suggested by the appearance of large red geological deposits. Photosynthesis produces oxygen as a by-product, and the gradual increase of photosynthesizing bacteria resulted in the buildup of oxygen in the atmosphere over hundreds of millions of years. Dissolved iron would oxidize in the presence of free oxygen, thus accounting for the "red beds." In the words of researcher William Schopf, "The Earth's oceans had been swept free of dissolved iron; lowly cyanobacteria—pond scum—had rusted

Figure 2.1
Timescale of Earth's History.
Courtesy of Alan Gishlick

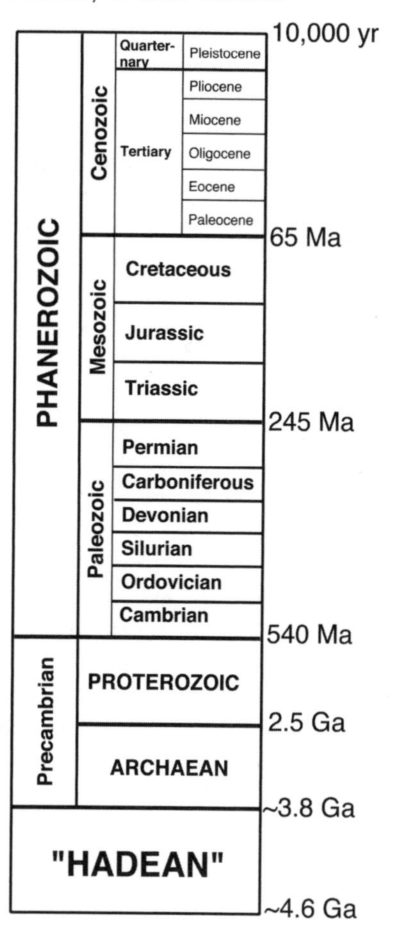

the world!" (Schopf 1992: 48). The increase of oxygen in the atmosphere resulted in a severe change in the environment: many organisms could not live in the new "poisonous" oxygenated environment. Others managed to survive and adapt.

Eukaryotic cells may have evolved as unnucleated cells incorporated by other cells within their membranes. Nucleated cells have structures called *organelles* within their cytoplasm which perform a variety of functions having to do with energy capture and use, cell division, predation, and other activities. Some of these structures, such as mitochondria and chloroplasts, have their own DNA. Similarities between the DNA of such organelles and that of some simple bacteria have supported the theory that early in evolution, the ancestors of eukaryotes absorbed certain bacteria and formed a cooperative or *symbiotic* relationship with them, the newcomers functioning to enhance performance of metabolism, predation, or some other task (Margulis 1993). The nucleus itself may also have been acquired in a similar fashion, from "recycled" parts obtained after the absorption of other bacteria. Evidence for these theories comes, of course, not from the fossil record but from inferences based on biochemical comparisons of living forms.

Once nucleated cells developed, sexual reproduction was not far behind. Sexual reproduction has the advantage of mixing up genetic information, allowing the organism to adapt to environmental change or challenge. Some researchers theorize that geological and atmospheric changes, together with the evolution of sexual reproduction, stimulated a burst of evolutionary activity during the late Precambrian period, about 900 million years ago, when the first multicellular animals, or *metazoans*, appear in the fossil record.

The Cambrian Explosion. As might be expected, the first evidence we have of metazoa is of simple creatures like sponges, jellyfish, echinoderms, and wormlike forms. But by the Cambrian, about 500 million years ago, we have rapid divergent evolution of invertebrate groups. Structures like body segmentation, a mouth and an anus, segmented appendages, and other "inventions" characterize new forms of animal life, some of which died out but many of which continue until the present day. These new body plans appear over a geologically sudden—if not biologically sudden—period of about 10–20 million years. Crustaceans, brachiopods, mollusks, and annelid worms, as well as representatives of other groups, appear in the Cambrian.

Evolutionary biologists are studying how these groups are related to one another, and investigating whether they indeed have roots in the Precambrian period. In evolutionary biology, as in the other sciences, theory building depends on cross-checking ideas against different types of data. There are three basic types of data used to investigate the evolutionary relationships among the invertebrate groups: size and shape (morphological) comparisons among modern representatives of these groups, biochemical comparisons among modern descendants, and the fossil record. Largely because of problems of preservation of key fossils at key times—and the fact that evolution might have taken "only" tens of millions of years, an eye blink from the perspective of deep time—the fossil evidence currently does not illuminate links among most of the basic invertebrate groups. Nonetheless, much nonfossil research is being conducted to understand similarities and differences of living members of these groups, from which we may infer evolutionary relationships.

One particularly interesting area of research has to do with understanding the evolution and developmental biology (embryology) of organisms, a new field referred to as "evo-devo."

"Evo-devo." Advances in molecular biology have permitted developmental biologists to study the genetics behind the early stages of embryological development in many groups of animals. What they are discovering is astounding. It is apparent that very small changes in genes affecting early, basic structural development can cause major changes in body plans. For example, there is a group of genes operating very early in animal development that is responsible for determining the basic front-to-back, top-to-bottom, and side-to-side orientations of the body. Other early-acting genes also control such bodily components as segments and their number, and the production of structures such as legs, antennae, and wings. Major changes in body plan can come about through rather small changes in these early-acting genes. What is perhaps the most intriguing result of this

research is the discovery of identical or virtually identical early genes in groups as different as insects, worms, and vertebrates. Could some of the body plan differences of invertebrate groups be the result of changes in genes that act early in embryological development?

Probable evolutionary relationships among the invertebrate groups are being established through anatomy, molecular biology, and genetics, even if they haven't been established through the fossil record. One tantalizing connection is between chordates, the group to which vertebrates belong (see below), and echinoderms, the group to which starfish and sea cucumbers belong. Based on embryology, RNA, and morphology, it appears as if the group to which humans and other vertebrates belong shared a common ancestor hundreds of millions of years ago with these primitive invertebrates. Although adult echinoderms don't look anything like chordates, their larval forms are intriguingly similar to primitive chordates. There are also biochemical similarities in the way they utilize phosphates—but read on to find out more about chordates!

Vertebrate Evolution. Our species is among the vertebrates, creatures with a bony structure encircling our nerve cord. Vertebrates are included in a larger set of organisms called *chordates*. Although all vertebrates are chordates, not all chordates are vertebrates. The most primitive chordates look like stiff worms. A notochord or rod running along the back of the organism with a nerve cord running above it is characteristic of chordates. At some time in a chordate's life, it also has slits in the neck region (which become gills in many forms) and a tail. An example of a living chordate is a marine filter-feeding creature an inch or so long called amphioxus. To look at it, you wouldn't think it was very closely related to vertebrates, but it is. Amphioxus lacks vertebrae, but like vertebrates, it has a notochord, a dorsal nerve cord, a mouth, an anus, and a tail. Like vertebrates, it is the same on the right side of the body as it is on the left (that is, it is bilaterally symmetrical), and it has some other similarities in the circulatory system and muscle system that are structurally similar to vertebrates. It is probably fairly similar to an early chordate, but because it has been around the planet for a long time, it has evolved as well. Still, it preserves the diagnostic features of chordates in a relatively simple form (Figure 2.2).

Figure 2.2
Amphioxus shows the basic "body plan" of chordates in having a mouth, an anus, a tail, a notochord, and a dorsal nerve chord. Courtesy of Janet Dreyer

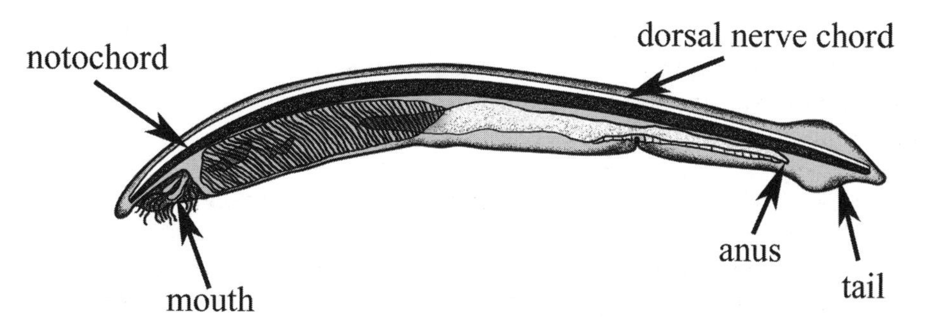

Now that you have some idea what a primitive chordate was like, let's return to the fact that larval forms of echinoderms had similarities to primitive chordates. Unlike adult echinoderms, which are radially symmetrical like starfish, echinoderm larval forms are bilaterally symmetrical like chordates. Embryologically, echinoderms and chordates form the anus and mouth from the same precursors, and they have a number of other developmental similarities as well. One hypothesis for chordate origins is that the larval form of an early echinoderm could have become sexually mature without "growing up"—going through the full metamorphosis to an adult. This phenomenon is uncommon, but it is not unknown. It occurs in salamanders such as the axolotl, for example.

Amphioxus is iconographic in biological circles. There aren't very many evolution songs (there are far more antievolution songs!), but one that many biologists learn is sung to the tune of "It's a Long Way to Tipperary":

Amphioxus
by Phillip H. Pope

(*chorus*)
It's a long way from Amphioxus
It's a long way to us
It's a long way from Amphioxus
To the meanest human cuss;
Goodbye fins and gill slits,
Hello lungs and hair,
It's a long long way from Amphioxus,
But we come from there!

(*verse*)
A fishlike thing appeared among the annelids one day
It hadn't any parapods nor cetae to display
It hadn't any eyes or jaws or ventral nervous cord,
But it had a lot of gill slits, and it had a notochord.

Well, it wasn't much to look at, and it scarce knew how to swim
And Nerius was very sure it hadn't come from him.
The mollusks wouldn't own it, and the arthropods got sore
So the poor thing had to burrow in the sand along the shore.

It burrowed in the sand before it grabbed in with its tail
And said, "Gill slits and myotomes are all to no avail
I've grown some metapleural folds and sport an oral hood
But all these fine new characters don't do me any good!"

He soaked a while down in the sand without a bit of pep
Then he stiffened up his notochord and said, "I'll beat 'em yet."
"They laugh and show their ignorance, but I don't mind their jeers
Just wait until they see me in a hundred million years!

"My notochord shall stiffen to a chain of vertebree,
As fins my metaplural folds will agitate the sea
My tiny dorsal nervous cord will be a mighty brain
And vertebrates will dominate the animal domain!"

In the Middle Cambrian is a small fossil called *Pikaia*, which is thought to be a primitive chordate because it looks rather like amphioxus (Figure 2.3). A new marine fossil discovered in the Late Cambrian Chengiang beds of China might even be a primitive vertebrate. Although *Haikouella* swam, it certainly didn't look much like a fish as we think of them today; it more resembled a glorified Amphioxus (Figure 2.4). From such primitive aquatic chordates as these eventually arose primitive jawless fish, then sharks and modern fish, and eventually the first land vertebrates: tetrapods (four-footed). These in turn became the ancestors of the other great groups of land animals, reptiles and mammals. Later, more detail will be provided about the evolution of many of these groups. But it is worthwhile to present four basic principles of biological evolution to keep in mind as you read the rest of the book. These are natural selection, adaptation, adaptive radiation, and speciation.

Figure 2.3
Pikaia, a Middle Cambrian fossil, shows some characteristics of primitive chordates. Courtesy of Janet Dreyer

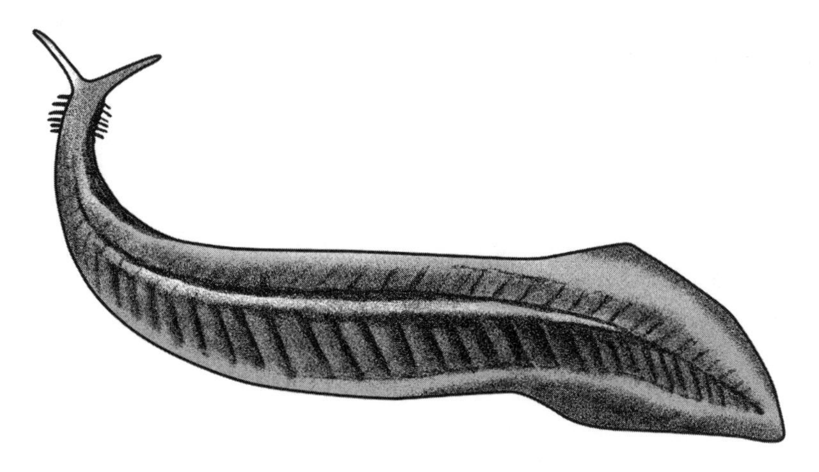

Figure 2.4
Haikouella, a Late Cambrian marine fossil, may be a primitive vertebrate. Courtesy of Janet Dreyer

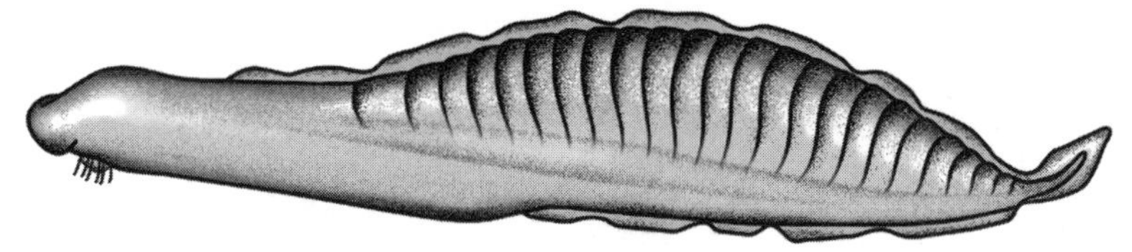

PRINCIPLES OF BIOLOGICAL EVOLUTION

Natural Selection and Adaptation

Natural selection is the term given by Charles Darwin to what he—and almost all modern evolutionary biologists—considered to be the most powerful force of evolutionary change. In fact, the thesis that evolution is primarily driven by natural selection is sometimes called *Darwinism*. Unfortunately, many people misapply the term to refer to the concept of descent with modification itself, which is erroneous. Natural selection is not the same as *evolution*. As discussed in chapter 1, there is a conceptual difference between a phenomenon and the mechanisms or processes that bring it about.

When Darwin's friend T. H. Huxley learned of the concept of natural selection, he said, "How extremely stupid not to have thought of that!" (Huxley 1888), so obvious did the principle seem to him—after it was formulated. And indeed, it is a very basic, very powerful, idea. The philosopher Daniel Dennett has called natural selection "the single best idea anyone has ever had" (Dennett 1995: 21). Because of its generality, natural selection is widely found not only in nature but increasingly in engineering, computer programming, the design of new drugs, and other applications.

The principle is simple: generate a variety of possible solutions, and then pick the one that works best for the problem at hand. The first solution is not necessarily the optimal one—in fact, natural selection rarely results in a good solution to a problem in one pass. But repeated iterations of randomly generated solutions, combined with selection of the characteristics that meet the necessary criteria, result in a series of solutions that more closely approximate a good solution. Engineers attempting to design new airplane wings have used natural selection approaches; molecular biologists trying to develop new drugs have also used the approach (Felton 2000). In living things, the "problem at hand," most broadly conceived, is survival and reproduction—passing on genes to the next generation. More narrowly, the "problem at hand" might be withstanding a parasite, finding a nesting site, being able to attract a mate, or being able to eat bigger seeds than usual when a drought reduces the number of small seeds. What is selected for depends on what, in the organism's particular circumstances, will be conducive to its survival and reproduction. The "variety of possible solutions" consists of genetically based variations which allow the organism to solve the problem.

Variation among members of a species is essential to natural selection, and fortunately it is common in sexually reproducing organisms. Some of these variations are obvious to us, such as differences in size, shape, or color. Other variations are invisible, such as genetically based biochemical and molecular differences that may be related to disease or parasite resistance, or the ability to digest certain foods. If the environment of a group of plants or animals presents a challenge—say heat, aridity, a shortage of hiding places, or a new predator—the individuals who just happen to have the genetic characteristics allowing them to survive longer and reproduce in that environment are going to be the ones most likely to pass on their genes to the next generation. The genes of these individuals—including those that better suit them to the environment—will increase in proportion to those of other individuals as the population reproduces itself generation after generation. The environment *naturally selects* those individuals with the characteristics that provide for a higher probability of survival, and thus those characteristics will tend to increase over time.

So the essence of natural selection is genetic variation within a population, an environmental condition that favors some of these variations more than others, and differential reproduction of the individuals who happen to have the favored variations.

A classic example of natural selection followed the introduction of rabbits into Australia—an island continent where rabbits were not native. In 1859, an English immigrant, Thomas Austin, released 12 pairs of rabbits so that he could go rabbit hunting. Unfortunately, except for the wedge-tailed eagle, a few large hawks, and dingoes (wild dogs)—and human hunters—rabbits have no natural enemies on Australia, and they reproduced like, well, rabbits. Within a few years, the rabbit population had expanded to such a large number that rabbits became a major pest, competing for grass with cattle, other domestic animals, and native Australian wildlife. Regions of the Australian outback that were infested with rabbits became virtual dustbowls as the little rodent herbivores nibbled down anything that was green. How could rabbit numbers be controlled?

Officials in Australia decided to import a virus from Great Britain that was fatal to rabbits but which was not known to be hazardous to native Australian mammals. The virus produced myxomatosis, or rabbit fever, which causes death fairly rapidly. It is spread from rabbit to rabbit by fleas or other blood-sucking insects. The virus first was applied to a test population of rabbits in 1950. Results were extremely gratifying: in some areas the count of rabbits decreased from 5,000 to 50 within six weeks. However, not all the rabbits were killed; some survived to reproduce. When the rabbit population rebounded, myxomatosis virus was reintroduced, but the positive effects of the first application were not repeated: many rabbits were killed, of course, but a larger percentage survived. Eventually, myxomatosis virus no longer proved effective in reducing the rabbit population. Subsequently, Australians have resorted to thousands of miles of rabbit-proof fencing to try to keep the rabbits out of at least some parts of the country.

How is this an example of natural selection? Consider how the three requirements outlined for natural selection were met:

1. Variation: The Australian rabbit population consisted of individuals that varied genetically in their ability to withstand the virus causing myxomatosis.
2. Environmental condition: Myxomatosis virus was introduced into the environment, making some of the variations naturally present in the population of rabbits more "valuable" than others.
3. Differential reproduction: Rabbits that happened to have variations allowing them to survive the viral disease reproduced more than others, leaving more copies of their genes in future generations. Eventually the population of Australian rabbits consisted of individuals that were more likely to have the beneficial variation. When myxomatosis virus again was introduced into the environment, fewer rabbits were killed.

Natural selection involves *adaptation*: having characteristics that allow an organism to survive and reproduce in its environment. Which characteristics increase or decrease in the population through time depends on the value of the characteristic, and that depends on the particular environment—it's not "one size fits all." Because environments can change, it is difficult to precisely predict what characteristics will increase or decrease, though general predictions can be made. (No one would predict that natural selection would produce naked mole rats in the Arctic, for example.) As a result, natural selection is sometimes defined as *adaptive differential reproduction*. It is *differential reproduction* because some individuals reproduce more or less than others. It is *adaptive* because the reason for the differential in reproduction has to do with a value that a trait of set of traits has in a particular environment.

Natural Selection and Chance. The myxomatosis example illustrates two important aspects of natural selection: it is dependent on the genetic variation present in the population, and on the "value" of some of the genes in the population. Some individual rabbits just happened to have the genetically based resistance to myxomatosis virus even before the virus was introduced; the ability to tolerate the virus wasn't generated by the "need" to survive under tough circumstances. It is a matter of chance which particular rabbits were lucky enough to have the set of genes conferring resistance. So is it correct to say that natural selection is a chance process?

Quite the contrary. Natural selection is the opposite of chance. It is *adaptive* differential reproduction: the individuals that survive to pass on their genes do so because they have genes that are helpful (or at least not negative) in a particular environment. Indeed, there are chance aspects to the production of genetic variability in a population: Mendel's laws of genetic recombination are, after all, based on probability. However, natural selection itself is far from being a chance process. If indeed evolution is driven primarily by natural selection, then it is not the result of chance. Now, during the course of a species' evolution, unusual things may happen that are outside of anything genetics or adaptation can affect, such as a mass extinction caused by an asteroid striking Earth, but such events—though they may be dramatic—are exceedingly rare. Such contingencies do not make evolution a chance phenomenon any more than your life is governed by chance because there is a 1 in 2.8 million chance of your being struck by lightning.

Natural Selection and Perfection of Adaptation. The first batch of Australian rabbits to be exposed to myxomatosis virus died in droves, though some survived to reproduce. Why weren't the offspring of these surviving rabbits completely resistant to the disease? A lot of *them* died, too, though a smaller proportion compared to those in the parent's generation. This is because natural selection usually does not result in perfectly adapted structures or individuals. There are several reasons for this, and one has to do with the genetic basis of heredity.

Genes are the elements that control the traits of an organism. They are located on chromosomes, in the cells of organisms. Because chromosomes are paired, genes also come in pairs, and for some traits, the two genes are identical. For mammals, genes that code for a four-chambered heart do not vary—or at least if there are any variants, the organisms that have them don't survive. But many genetic features do vary from individual to individual. Variation can be produced when the two genes of a pair differ, as they do for many traits. Some traits (perhaps most) are influenced by more than one gene, and similarly, one gene may have more than one effect. The nature of the genetic material and how it behaves is a major source of variation in each generation.

The rabbits that survived the first application of myxomatosis bred with one another, and because of genetic recombination, some offspring were produced that had myxomatosis resistance, and others were produced that lacked the adaptation. These latter were the ones who died in the second round when exposed to the virus. Back in Darwin's day, a contemporary of his invented a "sound bite" for natural selection: he called it "survival of the fittest," with "fit" meaning best adapted—not necessarily the biggest and strongest. Correctly understood, though, natural selection is survival of the fit *enough*. It is not, in fact, only the individuals who are most perfectly suited to the environment that survive; reproduction, after all, is a matter of degree, with some rabbits (or humans or spiders or oak trees) reproducing at higher than the average rate, and some at lower than the average rate. As long as an individual reproduces at all, though, it is *fit*, even if some are fitter than others.

Furthermore, just as there is selection within the rabbit population for resistance to the virus, so there is selection among the viruses! Viruses need to get copies of themselves into the next generation, and they can do this only in the body of a live rabbit. If the infected rabbit dies too quickly, the virus doesn't get a chance to spread. Viruses that are too virulent tend to be selected against, just as the rabbits that are too susceptible will also be selected against. The result is an evolutionary contest between host and pathogen—reducing the probability that the rabbit species will ever be fully free of the virus.

Another reason why natural selection doesn't result in perfection of adaptation is that once there has been any evolution at all (and there has been considerable animal evolution since the appearance of the first metazoa), there are constraints on the direction in which evolution can go. As discussed elsewhere, if a vertebrate forelimb is shaped for running, it would not be expected to become a wing at a later time; that is one kind of constraint. Another constraint is that natural selection has to work with structures and variations that are available, regardless of what sort of architecture could best do the job. If you needed a guitar but all you had was a toilet seat, you *could* make a sort of guitar by running strings across the opening, but it wouldn't be a perfect design. The process of natural selection works more like a tinkerer than an engineer (Monod 1971).

Evolution and Tinkering. Some builders are engineers and some are tinkerers, and the way they go about constructing something differs quite a lot. An engineering approach to building a swing for little Charlie is to measure the distance from the tree branch to a few feet off the ground; go to the hardware store to buy some chain, hardware, and a piece of wood for the bench; and assemble the parts, using the appropriate tools: measuring devices, a drill, a screwdriver, screws, a saw, sandpaper, and paint. Charlie ends up with a really nice, sturdy swing that avoids the "down will come baby, cradle, and all" problem and that won't give slivers to his little backside when he sits on it. A tinkerer, on the other hand, building a swing for little Mary, might look around the garage for a piece of rope, throw it over the branch to see if it is long enough, and tie it around an old tire. Little Mary has a swing, but it isn't quite the same as Charlie's. It gets the job done, but it certainly isn't an optimal design: the rope may suspend little Mary too far off the ground for her to be able to use the swing without someone to help her get into it; the rope may be frayed and break; the swing may be suspended too close to the trunk, so Mary careens into it—you get the idea. The tinkering situation, where a structural problem is solved by taking something extant that can be bent, cut, hammered, twisted, or manipulated into something that more or less works, however crudely, mirrors the process of evolution much more than do the precise procedures of an engineer. Nature is full of structures that work quite well—but it also is full of structures that just barely work, or that, if one were to imagine designing from scratch, one would certainly not have chosen the particular modification that natural selection did.

Several articles by Stephen Jay Gould have discussed the seemingly peculiar ways some organisms get some particular job done. An angler fish has a clever "lure" resembling a wormlike creature that it waves at smaller fish to attract them close enough to eat. The lure springs from its forehead and is actually a modified dorsal fin spine (Gould 1980a). During embryological development, the panda's wrist bone is converted into a sixth digit, forming a grasping hand out of the normal five fingers of a bear paw, plus a "thumb" that is jury-rigged out of a modifiable bone (Gould 1980b). Like a tinkerer's project, it gets the job done, even if it isn't a great design. And after all, it's survival of the fit *enough*.

Natural selection is usually viewed as a mechanism that works on a population or sometimes a species to produce adaptations. Natural selection can also bring about adaptation on a very large scale through *adaptive radiation*.

Adaptive Radiation

To be fruitful and multiply, all living things have to acquire energy (through photosynthesis, or from consuming other living things), they have to avoid predation and illness, and they have to reproduce. As is clear from the study of natural history, there are many different ways that organisms manage to perform these tasks, reflecting both the variety of environments on Earth and the variety of living things. Any environment—marine, terrestrial, arboreal, aerial, subterranean—contains many *ecological niches* that provide means that living things use to make a living. The principle of *adaptive radiation* helps to explain how niches get filled.

The geological record reveals many examples of the opening of a new environment and its subsequent occupation by living things. Island environments such as the Hawaiian Islands, the Galapagos Islands, Madagascar, and Australia show this especially well. The Hawaiian archipelago was formed as lava erupted from undersea volcanoes, and what we see as islands actually are the tips of volcanic mountains. Erosion produced soils, and land plants—their seeds or spores blown or washed in—subsequently colonized the islands. Eventually land animals reached the islands as well. Birds, insects, and a species of bat were blown to Hawaii or rafted there from other Pacific islands on chunks of land torn off by huge storms.

The Hawaiian honeycreepers are a group of approximately 23 species of brightly colored birds, ranging from four to eight inches long. They have been extensively studied by ornithologists and are shown to be very closely related. Even though they are closely related, honeycreeper species vary quite a bit from one another and occupy many different ecological niches. Some are insectivorous, some suck nectar from flowers, others are adapted to eating different kinds of seeds—one variety has even evolved to exploit a woodpecker-like niche. The best explanation for the similarity of honeycreepers in Hawaii is that they are all descended from a common ancestor. The best explanation for the diversity of these birds is that the descendants of this common ancestor diverged into many subgroups over time, as they became adapted to new, open ecological niches. Honeycreepers are, in fact, a good example of the principle of adaptive radiation, where one or a few individual animals arrive in a new environment that has empty ecological niches, and their descendants are selected to quickly evolve the characteristics needed to exploit these niches. Lemurs on Madagascar, finches on the Galapagos Islands, and the variety of marsupial mammals in Australia and prehistoric South America are other examples of adaptive radiation.

A major adaptive radiation occurred in the Ordovician period (about 430 million years ago), when plants developed protections against drying out, protections against ultraviolet radiation, vascular tissue to support erect stems, and other adaptations allowing life out of water (Richardson 1992). It was then that the dry land could be colonized by plants. The number of free niches allowed plants to radiate into a huge number of ways of life. The movement of plants from aquatic environments onto land was truly an Earth-changing event. Another major adaptive radiation occurred when vertebrates evolved adaptations (lungs and legs) permitting a similar movement onto land. These early tetrapods radiated into amphibians, reptiles, and mammals. During the late Cretaceous and early Cenozoic, about 65 million years ago, mammals began adaptively radiating after the demise of the dinosaurs opened up new ecological niches for them. Mammals moved into gnawing niches (rodents), a variety of grazing and browsing niches (hoofed quadrupeds, the artiodactyls and perissodactyls), insect-eating niches (insectivores and primates), and meat-eating niches (carnivores). Over time, subniches were occupied: some carnivores stalk their prey (lions, saber-toothed cats), others run it down (cheetahs, wolves); some (lions, wolves, hyenas) hunt large-bodied prey, some (foxes, bobcats) hunt small prey.

If a particular adaptive shift requires extensive changes, such as greatly increasing or reducing the size or number of parts of the body, the tendency is for that change to occur early in the evolution of the lineage rather than later. Though not a hard-and-fast rule, it follows logically from natural selection that the greatest potential for evolutionary change will occur before specializations of size or shape take place. Early in evolutionary history, the morphology of a major group tends to be more generalized, but as adaptive radiation takes place, structures are selected for to enable the organisms to adapt better to their environments. In most cases, these adaptations *constrain* or limit future evolution in some ways. The forelimbs of perch are committed to propelling them through the water and are specialized for this purpose; they will not become grasping hands.

We and all other land vertebrates have four limbs. Why? We tend to think of four limbs as being "normal," yet there are other ways to move bodies around. Insects have six legs and spiders have eight, and these groups of animals have been very successful in diversifying into many varieties and are represented in great numbers all over the world. So there's nothing especially superior about having four limbs, although apparently, since no organism has evolved wheels for locomotion, two or more legs apparently work better. But all land vertebrates have four limbs rather than six or eight because reptiles, birds, and mammals are descended from early four-legged creatures. These first land vertebrates had four legs because the swimming vertebrates that gave rise to them had two fins in front and two in back. The number of legs in land vertebrates was *constrained* because of the number of legs of their aquatic ancestors. Imagine what life on Earth would have looked like if the first aquatic vertebrates had had six fins! Might there have been more ecological niches for land creatures to move into? It certainly would have made sports more interesting if human beings had four feet to kick balls with—or four hands to swing bats or rackets.

We see many examples of constraints on evolution; mammalian evolution provides another example. After the demise of the dinosaurs, mammals began to radiate into niches that had previously been occupied by the varieties of dinosaurs. As suggested by the shape of their teeth, mammals of the late Cretaceous and early Paleocene were small, mostly undifferentiated creatures occupying a variety of insectivorous, gnawing, and seed-eating niches that dinosaurs were not exploiting. As new niches became available, these stem mammals quickly diverged into basic mammalian body plans: the two kinds of hoofed mammals, the carnivores, bats, insectivores, primates, rodents, sloths, and so on. Once a lineage developed (carnivores, for example), it radiated within the basic pattern to produce a variety of different forms (cats, dogs, bears, raccoons, etc.) in many sizes and shapes, all having inherited basic dental and skeletal traits from the early carnivore ancestor. Once a lineage is "committed" to a basic way of life, it is rare indeed for a major adaptive shift of the same degree to take place. Although both horses and bats are descended from generalized quadrupedal early mammalian ancestors, the bones in a horse's forelimbs have been modified for swift running: some bones have been greatly elongated, others have been lost completely, and others have been reshaped. A bat has the same basic bones in its forelimb, but they have been greatly modified in other ways: some bones have been elongated, others have been lost, and yet others have been reshaped for flight (Figure 2.5).

Humans belong in the primate group of mammals, and primates are characterized by relatively fewer skeletal changes than have occurred in other mammal lineages. A primate doesn't have the extensive remodeling of the forelimb and hand that resulted in a bat's wing or a horse's hoof. We primates have a relatively basic "four on the floor" quadruped limb pattern of one bone close to the body (the femur in the leg and the humerus in the arm), two bones next to this one (leg: tibia and fibula; arm: radius and ulna), a group of small bones after this (leg: tarsal or ankle bones; arm: carpal or wrist bones), and a fanlike spray of small bones at the end of the limb (leg: metatarsals and toe bones; arm: metacarpals and finger bones) (Figure 2.5). Most primates locomote using four limbs; we human primates have taken this quadrupedal pattern and tipped it back so that all our weight is borne on the hind limbs (and not too successfully, as witnessed by hernias and the knee and lower back problems that plague our species). Being bipedal, though, meant we did not have to use our hands for locomotion, and they were thus freed for other purposes, like carrying things and making tools. Fortunately for human beings, our early primate ancestors did not evolve to have specialized appendages like those of horses or bats.

Figure 2.5
Vertebrate forelimbs all contain the same bones although these bones have evolved over time for different locomotor purposes: running, swimming, flying, grasping, etc. Courtesy of Janet Dreyer

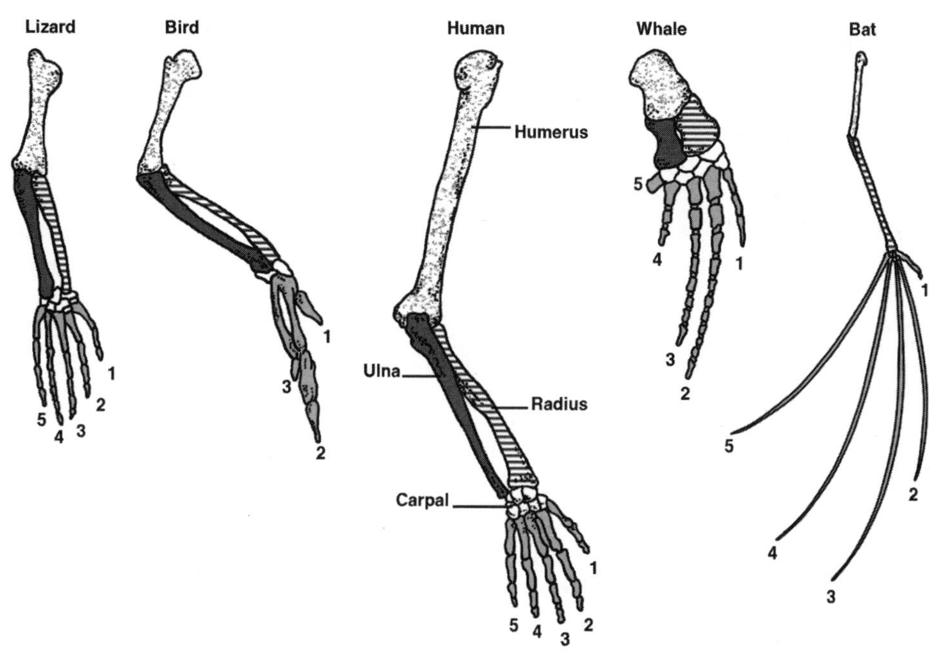

Which is better—to be generalized or specialized? It's impossible to say without knowing more about the environment or niche in which a species lives. Specialized organisms may do very well by being better able to exploit a resource than possible competitors; yet generalized organisms have an advantage in being able to adapt to a new environmental challenge.

Speciation

A *species* is composed of all the individuals that can exchange genes with one another. Some species are composed of very few individuals located in a restricted area, and others have millions of members spread out over large areas of the world. Some plant species are restricted to small areas of rain-forest habitat, while rats and humans live on literally every continent. It is more likely that an individual will mate with another individual living close by than farther away, and as a result, most species can be divided into smaller populations. Sometimes geographical factors, such as rivers or mountains or temperature gradients in different depths of water, will naturally carve species into populations.

Because of geographical differences among populations, natural selection tends to result in populations varying from one another. A typical widespread species may be divided into many differing populations. As long as they exchange genes at least at intervals, populations are likely to remain part of the same species. But how do new species form? New species form when members of a population or subdivision of a species no longer are able to exchange genes with the rest of the species. This is more likely to happen at the edges of the species range than in the center. We can say that *speciation* has occurred when a population becomes *reproductively isolated* from the rest of the species.

If a population at the end of the geographic range of a species is cut off from the rest of the species, through time it may become different from other populations. Perhaps natural selection is operating differently in its environment than it is in the rest of the species range, or perhaps the population has a somewhat different set of genes than other populations of the species. Just by the rules of probability, a small, peripheral population is not likely to have all the variants of genes that are present in the whole species, which might result in its future evolution taking a different turn.

No longer exchanging genes with other populations of the species, and diverging genetically through time from them, members of a peripheral, isolated population might reach the stage where, were they to have the opportunity to mate with a member of the "parent" species, they would not be able to produce offspring. *Isolating mechanisms*, most of which are genetic but some of which are behavioral, can arise to prevent reproduction between organisms from different populations. Some isolating mechanisms prevent two individuals from mating at all: in some insects, for example, the sexual parts of males and females of related species are so different in shape or size that copulation can't take place. Other isolating mechanisms come into effect when sperm and egg cannot fuse for biochemical or structural reasons. An isolating mechanism could take the form of the prevention of implantation of the egg or of disruption of the growth of the embryo after a few divisions. Or the isolating mechanism could kick in later: mules, which result from crossing horses and donkeys, are healthy but sterile. Donkey genes thus are inhibited from entering into the horse species, and vice versa. When members of two groups aren't able to share genes because of isolating mechanisms, we can say that speciation between them has occurred. (Outside of the laboratory, it may be difficult to determine whether two species that no longer live in the same environment are reproductively isolated.)

The new species would of course be very similar to the old one—in fact, it might not be possible to tell them apart. Over time, though, if the new species manages successfully to adapt to its environment, it might also expand and bud off new species, which would be yet more different from the parent—now "grandparent"—species. This branching and splitting has, through time, given us the variety of species that we see today.

We can see this process of speciation operating today. Speciation in the wild usually takes place too slowly to be observed during the lifetime of any single individual, but there have been demonstrations of speciation under laboratory conditions. The geneticist Dobzhansky and his colleagues isolated a strain of Venezuelan fruit flies and bred it for several years. This strain of flies eventually reached a point of differentiation where it was no longer able to reproduce with other Venezuelan strains with which it had formerly been fertile. Speciation had occurred (Dobzhansky and Pavlovsky 1971).

Though not observed directly, good inferential evidence for speciation can be obtained from environments that we know were only recently colonized. The Hawaiian and Galapagos islands have been formed within the last few million years from undersea volcanoes and have acquired their plants and animals from elsewhere. The Galapagos flora and fauna are derived from South America, whereas the native Hawaiian flora and fauna are more similar to those of the Pacific islands, which in turn are derived mostly from Asia. But Hawaiian species are reproductively isolated from their mainland counterparts.

One of the most dramatic examples of speciation took place among cichlid fish in the East African great lakes: Lake Victoria, Lake Malawi, and Lake Tanganyika. Geological evidence indicates that about 25,000 years ago, Lake Tanganyika underwent a drying spell that divided the lake into three separate basins. Perhaps as a result of this and similar episodes, the cichlid fish that had entered the lake from adjacent rivers and streams underwent explosive adaptive radiation. There are at least 175 species of cichlid fish found in Lake Tanganyika and nowhere else. Similar speciation events took place in Lake Victoria and Lake Malawi—only over shorter periods of time (Goldschmidt 1996). Large lakes like these can be watery versions of an island: interesting biological things can go on!

Occasionally speciation can take place very quickly. The London subway, or "Tube," was built during the 1880s. At that time, some mosquitoes found their way into the miles of tunnels, and they successfully bred in the warm air and intermittent puddles—probably several times per year. Because they were isolated from surface mosquitoes, differences that cropped up among them would not have been shared with their relatives above, and vice versa. In the late 1990s, it was discovered that the Tube mosquitoes were a different species from the surface species. One major, if unfortunate, difference is that the surface mosquitoes, *Culex pipiens*, bite birds, whereas the related Tube species, called *Culex molestus*, has shifted its predation to people. What is surprising about this discovery is that it shows that at least among rapidly breeding insects like mosquitoes, speciation does not require thousands of years but can occur within a century (Bryne and Nichols 1999). Natural selection, adaptation, adaptive radiation, and speciation—these are the major principles that help us explain the pattern and understand the process of evolution.

DID MAN EVOLVE FROM MONKEYS?

So, to end with the question we began with: Did man evolve from monkeys? No. The concept of biological evolution, that living things shared common ancestry, implies that human beings did not descend from monkeys, but shared a common ancestor with them, and shared a common ancestor farther back in time with other mammals, and farther back in time with reptiles, and farther back in time with fish, and farther back in time with worms, and farther back in time with petunias. We are not descended from petunias, worms, fishes, or monkeys, but we shared common ancestors with all of these creatures, some more recently than others. The inference of common ancestry helps us make sense of biological variation. Humans are more similar to monkeys than we are to dogs because we shared a common ancestor with monkeys more recently than we shared a common ancestor with dogs. Humans, dogs, and monkeys are more similar to one another (they are all mammals) than they are to salamanders, because the species that provided the common ancestor of all mammals lived more recently than the species providing the common ancestors of salamanders and mammals. This historical branching relationship of species through time allows us to group species into categories such as "primates," "mammals," and "vertebrates," which allows us to hypothesize about other relationships. Indeed, the theory of evolution, as one famous geneticist put it, is what "makes sense" of biology. "Seen in the light of evolution, biology is, perhaps, the most satisfying science. Without that light it becomes a pile of sundry facts, some of them more or less interesting, but making no comprehensible whole" (Dobzhansky 1973: 129).

REFERENCES CITED

Bryne, Katharine, and Richard A. Nichols. 1999. *Culex pipiens* in London Underground Tunnels: Differentiation Between Surface and Subterranean Populations. *Heredity* 82: 7–15.

Dennett, Daniel C. 1995. *Darwin's Dangerous Idea.* New York: Simon and Schuster.

Dobzhansky, Theodosius. 1973. Nothing in Biology Makes Sense Except in the Light of Evolution. *American Biology Teacher* 25: 125–129.

Dobzhansky, Theodosius, and O. Pavlovsky. 1971. Experimentally Caused Incipient Species of *Drosophila. Nature* 230: 289–292.

Felton, Michael J. 2000. Survival of the Fittest in Drug Design. *Drug* 3 (9): 49–50, 53–54.

Goldschmidt, Tijs. 1996. *Darwin's Dreampond: Drama in Lake Victoria*, translated by S. Marx-MacDonald. Cambridge, MA: MIT Press.

Gould, Stephen Jay. 1980a. Double Trouble. In Gould's *The Panda's Thumb: More Reflections on Natural History*. New York: Norton.

Gould, Stephen Jay. 1980b. The Panda's Thumb. In Gould's *The Panda's Thumb: More Reflections on Natural History*. New York: Norton.

Gould, Stephen Jay. 1987. *Time's Arrow, Time's Cycle: Myth and Metaphor in the Discovery of Geological Time*. Cambridge: Harvard University Press.

Hazen, Robert M., Timothy R. Filley, Glenn A. Goodfriend. 2001. Selective adsorption of L- and D-amino acids on calcite: Implications for biochemical homochirality. *Proceedings of the National Academy of Sciences* 98 (10): 5487–5490.

Huxley, Thomas Henry. 1888. On the Reception of the "Origin of Species." In *The Life and Letters of Charles Darwin*, edited by F. Darwin. London: Thomas Murray.

Joyce, Gerald F. 1991. The Rise and Fall of the RNA World. *The New Biologist* 3: 399–407.

Lewis, Ricki. 1997. Scientists Debate RNA's Role at Beginning of Life on Earth. *The Scientist* 11 (7): 11.

Margulis, Lynn. 1993. *Symbiosis in Cell Evolution*, 2nd ed. New York: Freeman.

Miller, Stanley L. 1992. The Prebiotic Synthesis of Organic Compounds as a Step Towards the Origin of Life. In *Major Events in the History of Life*, edited by W. J. Schopf. Boston: Jones and Bartlett.

Monod, Jacques. 1971. *Chance and Necessity*. New York: Knopf.

Nelson, Kevin E., Matthew Levy, and Stanley L. Miller. 2000. Peptide Nucleic Acids Rather Than RNA May Have Been the First Genetic Molecule. *Proceedings of the National Academy of Sciences* 97 (8): 3868–3871.

Richardson, John B. 1992. Origin and Evolution of the Earliest Land Plants. In *Major Events in the History of Life*, edited by J. W. Schopf. Boston: Jones and Bartlett.
Rode, Bernd Michael. 1999. Peptides and the Origin of Life. *Peptides* 20: 773–786.
Schopf, J. William, ed. 1992. *Major Events in the History of Life*. Boston: Jones and Bartlett.
Silk, Joseph. 1994. *A Short History of the Universe*. New York: Scientific American Library.
Strickberger, Monroe. 1996. *Evolution*, 2nd ed. Sudbury, MA: Jones and Bartlett.

CHAPTER 3

Beliefs: Religion, Creationism, and Naturalism

Because the methodology of science works so well, you will find people from every nation, religion, and culture using it. Science is recognized internationally as the best way to find out about the natural world. But the *natural* world is not the only thing that human beings ask questions about, are concerned about, or think about. In fact, in every known human society, from the most complex urban civilization to the simplest community of hunter-gatherers, most people believe that there is a universe or world or *something* beyond or other than this material one, which is populated by gods, spirits, ancestors, or other nonmaterial beings. Science doesn't tell us anything about this world; this transcendent world is the provenance of religion.

RELIGION

Americans are most familiar with the Middle Eastern monotheistic traditions of the Jews, Christians, and Muslims. They are called "Abrahamic" religions because the patriarch Abraham is revered by all three, and their practitioners worship a single God who reveals Himself through sacred writings (the Torah, the Bible, and the Koran). All human societies have religious beliefs, however, and it is important not to let our understanding of a human universal such as religion be limited only to that which is familiar. To understand religion, one must look at, as well as beyond, the great Abrahamic religions.

All human societies have some belief system that can be called "religion." Some of these are believed in by hundreds of millions of people, such as Christianity, Islam, Confucianism, and Hinduism, whereas others are believed in by tribal groups whose numbers are reckoned in thousands or even fewer. With such a disparity of beliefs, can we find any commonalities?

One thing all religions appear to have in common is a belief in something beyond the material world, an Ultimate or Absolute or transcendent reality beyond the mundane. A sense of sacredness, awe, or mystery about this Beyond is common to religious beliefs and practices, and almost universal is the notion of spiritual (rather than mortal) beings that inhabit this realm and that have special powers. These include gods, witches, powerful spirits, and the like. Most religions, though not all, include the concept of life after death, and most include a component of worship—ritual behavior associated with these spiritual beliefs.

Intermediaries (such as priests and shamans) between people and the spiritual world are often very powerful and authoritative. Commonly there are special places for worship (temples, churches, holy sites) that are set apart from other sites (Stevens 1996). In virtually all religions, knowledge (about the supernatural; about where people, animals, and other natural objects came from; and about moral and ritual conduct) is obtained partly by revelation from supernatural sources. The gods of the Greeks revealed information through oracles, and the God of the Hebrews gave the Ten Commandments to Moses. Sometimes this revealed truth is recorded in texts that believers consider to be holy, such as the Koran of the Muslims, the Hindu Vedas, the Book of Mormon, or the New Testament of the Christians. Believers may dispute among themselves as to the proper interpretation of these holy texts.

How believers in a particular religion conceive of the Ultimate varies enormously, from views similar to the Christian personal God to the considerably more diffuse Hindu conception of Brahman, a generalized "spirit behind, beneath, and beyond the world of matter and energy" (Raman 1998–1999: 6). Even within Christianity, the concept of God varies widely from an anthropomorphic creator-God, such as that portrayed by Michelangelo on the ceiling of the Sistine Chapel, to a generalized force undergirding the universe that, although a source of awe, some Christians neither regard as a person nor pray to.

Human societies could not function without ethical systems—rules for behavior toward other people—and usually, though not universally, these systems are determined or at least strongly influenced by religion. In many human societies, it is believed that rules for behavior are divinely revealed, such as the Ten Commandments, which Christians, Jews, and Muslims believe were given by God to Moses. Others may ascribe the rules for proper behavior to directives from ancestors, and still others have no supernatural source for their rules but attribute the origin of such rules to custom alone.

RELIGION AND EXPLANATION

Although the primary function of religion is to mediate between people and the gods or forces beyond everyday existence, it may additionally provide explanations of the natural world. In many human societies, natural phenomena are frequently explained by reference to supernatural causation. The sun shines, or rain falls, but some sort of personal causation is involved in producing this effect. For example, the Brazilian Kuikuru "know it was the wind that blew the roof off a house, but they carry the search for explanation one step further and ask, 'Who sent the wind?'" A human or spirit personality "had to direct the natural force of the wind to produce its effect" (Carneiro 1983). Sickness, death, meteorological phenomena such as rain or tornadoes, the existence and location of mountains and other landforms, earthquakes, volcanoes, the passage of seasons, and the positions of the sun, stars, and planets also frequently have religiously based explanations. In fact, for most people living in tribal, nonindustrial settings, the natural world and the spiritual world are not divided but are blended, in contrast to the modern Western cultural view.

In earlier times in Western society, it was common for biblical statements about the natural world to be accepted as authoritative and for God to be viewed as the direct cause of natural events. If plague struck a community, or if a comet blazed across the sky, the event was attributed to the direct action of God, specially intervening in His creation. Gradually, though, some of these statements in the Bible were discarded as they were found to be inaccurate—for example, that Earth is a circle (reflecting early civilization's belief that the world was disk-shaped rather than spherical). Livestock breeders found that coat color in goats would not be affected by watering them at troughs in which peeled sticks had been placed (as claimed in Genesis 30), and thus the Bible came to be taken less as a source of information about the natural world, and more as a guide to understanding the relationship of man to God. St. Augustine, among other early church leaders, argued in the fourth and fifth centuries that it was bad theology to accept biblical statements about the natural world uncritically if such statements contradicted experience. He felt that too strict adherence to biblical literalism regarding statements about the natural world would diminish the credibility of proselytizers.

> Usually, even a non-Christian knows something about the earth, the heavens, and the other elements of this world, about the motion and orbit of the stars and even their size and relative positions, about the predictable eclipses of the sun and moon, the cycles of the years and the seasons, about the kinds of animals, shrubs, stones and so forth, and this knowledge he holds to as being certain from reason and experience.
>
> Now, it is a disgraceful and dangerous thing for an infidel to hear a Christian, while presumably giving the meaning of Holy Scripture, talking nonsense on these topics. . . . If they find a Christian mistaken in a field which they themselves know well, and hear him maintaining his foolish opinions about the Scriptures, how then are they going to believe those Scriptures in matters concerning the resurrection of the dead, the hope of eternal life, and the kingdom of heaven? How indeed, when they think that their pages are full of falsehoods on facts which they themselves have learnt from experience and the light of reason? (Saint Augustine 1982: 42–43)

During the seventeenth and eighteenth centuries, science was developing as a methodology of knowing about the natural world. "Natural philosophy," the study of nature, was regarded equally as a means to understand the mind of God and a means to understand the natural world. A considerable increase in knowledge about the natural world was obtained through the systematic methodology of science, in which natural phenomena were explained by being grouped under natural laws or theories. God was by no means ignored, but the focus was upon discovering the laws that God had created. Isaac Newton, for example, was a highly religious man who sought to discover the natural laws by which God governed the universe. He felt that a God who worked through His created natural laws was a God more worthy of awe and worship than one who constantly intervened to maintain the universe. To Newton, God was more awesome if He caused planets to orbit about the sun using gravity than if

He directly suspended them. Of course, as an omnipotent being, God *could* intervene at any time in the operation of the universe—miracles were possible—but it was not considered blasphemous to conclude that God acted through secondary causes (interpreted to be God's laws).

By the mid-nineteenth century, the success of science as a way of understanding the natural world was clear. It was *possible* to explain geological strata, for example, by reference to observable forces of deposition, erosion, vulcanism, and other processes, rather than having to rely upon the direct hand of God to have formed the layers. By the late nineteenth century, science was well on its way to avoiding even the occasional reliance upon God as immediate cause and to invoking only natural cause in explaining natural phenomena. This change in emphasis occurred not because of any animosity toward religion; rather, limiting science *only* to natural causes came about because it worked: a great deal was learned about the natural world by applying *materialist* (matter, energy, and their interaction) explanations.

Twentieth- and twenty-first-century scientists limit themselves to explaining natural phenomena using only natural causes for another practical reason: if a scientist is "allowed" to refer to God as a direct causal force, then there is no reason to continue looking for a natural explanation. Scientific explanation screeches to a halt. If there *were* a natural explanation, perhaps unknown or not yet able to be studied given technological limits or inadequate theory, then it would never be discovered if scientists, giving up in despair, invoked the supernatural. Scientists are quite used to saying "I don't know *yet*."

Perhaps the most important reason scientists restrict themselves to materialist explanations is that the methods of science are inadequate to test explanations involving supernatural forces. Recall that one of the hallmarks of science is the ability to hold some variables constant in order to be able to test the role of others. If indeed there is an omnipotent force that intervenes in the material world, by definition it is not possible to control for—to hold constant—such actions. As one wag put it, "You can't put God in a test tube"; and, one must add, you can't keep Him out of one, either. Such is the nature of omnipotence—by definition. As a result, scientists do not consider supernatural explanations as scientific. As a matter of fact, limiting scientific explanation to natural causes has been extraordinarily fruitful. In the spirit of the adage "If it ain't broke, don't fix it," scientists continue to seek explanations in natural processes when doing science, whether they are believers or nonbelievers in an omnipotent power.

A topic to which we will return at the end of the chapter concerns a difference between a rule of science and a philosophical view—between something called "methodological naturalism" and "philosophical naturalism." We have been discussing a rule of science that requires that scientific explanations use only material (matter, energy, and their interaction) cause; this is known as *methodological naturalism*. To go beyond methodological naturalism to claim that only natural causes exist—that is, that there is no God or, more generally, no supernatural entities—is *philosophical naturalism*. The two views are logically distinct because one can be a methodological naturalist but not accept naturalism as a philosophy. Many scientists who are theists are examples: in their scientific work they explain natural phenomena in terms of natural causes, even if in their personal lives they believe in God, and even that God may intervene in nature.

Christianity and many other religions rely at least in part upon truth revealed from God. When a revelation-based claim about the natural world is made, it may come into conflict with knowledge gained from experience—as Augustine described in the quote earlier in this chapter. What should a believer do when such conflicts arise? Such conflicts have arisen in the past, and continue to play an important role in the present-day creationism/evolution controversy.

A classic example of revealed truth's conflicting with scientific interpretation is the debate that took place in the seventeenth century regarding the relationship of Earth to the other planets and the sun. Traditionally, the Bible was interpreted as reflecting a geocentric, or Earth-centered, model of the universe. The sun and the other planets revolved around Earth. Early astronomers such as Copernicus and Galileo challenged the geocentric view, based on their empirical observations, inferences, and mathematical calculations, holding instead the heliocentric view that Earth and other planets revolved around the sun. The Catholic Church rejected these conclusions partly on scientific grounds, but primarily because heliocentrism contradicted the accepted interpretation of the Bible that Earth had to be the center of the universe. God had created humankind to worship Him, and, in turn, had made the whole universe for us. Because Earth was the place where human beings lived, logically it would be the center of the universe. Bible passages such as Joshua 10:12–13 reinforced this view. Joshua requests God to lengthen the day so his soldiers might win on the battlefield; God lengthens the day by stopping the sun, reflecting the geocentric model of the universe extant when the book of Joshua was set down. Although at one time, heliocentrism was considered blasphemous, today only a tiny fraction of Christians interpret the Bible as a geocentric document; for the vast majority of Christians, it is no longer necessary to interpret the Bible as presenting a geocentric worldview.

CREATIONISM

Just as with evolution, the word "creationism" has a broad and a narrow definition. Broadly, "creationism" refers to the idea that a supernatural force created. To Christians, Jews, and Muslims, this supernatural force is God; to people of other religions, it is other deities. The creative power may be unlimited, like that of the Hebrew God, or it may be restricted to the ability to affect certain parts of nature, such as heavenly bodies or certain kinds of living things.

The term "creationism" to many people connotes the theological doctrine of special creationism: that God created the universe essentially as we see it today, and that this universe has not changed appreciably since that creation event. Special creationism includes the idea that God created living things in their present forms, and it reflects a literalist view of the Bible. It is associated with an endeavor called "creation science," which includes the view that the universe is only 10,000 years old. Creation science and special creationism are discussed at greater length below.

It is important to define terms and use them consistently. In this book, the usual connotation of "creationism" will be the Christian view that God created directly.

Special creationism is the most familiar form of direct creationism, but some Christians view God as creating sequentially rather than all at once, as in special creationism. Later in this chapter readers will be introduced to a range of religious views about creationism and evolution which will help to make this definition clearer.

ORIGIN MYTHS

All people try to make sense of the world around them, and that includes speculating about the course of events that brought the world and its inhabitants to their present state. Stories of how things came to be are known as *origin myths*. They are tied to the broad definition of creationism.

Now, just as the word "theory" is used differently in science than it is used in casual conversation (see chapter 1), so the word "myth" is used in the anthropological study of cultures differently than the way we tend to use it informally. The common connotation of "myth" is something that is untrue, primitive, or superstitious—something that should be discounted. Yet when anthropologists talk of myths, it is to describe stories within a culture that symbolize what members of the culture hold to be most important. Although myths tend to be more common in nonliterate societies, they occur even in developed countries like our own. The "little engine that could," for example, expresses an important value in American culture: persevering in the face of adversity. The Horatio Alger myth of the poor but plucky youth who achieves success through hard work, pulling himself up by his bootstraps, is classically American. Both of these secular myths also express the American value of individualism—something quite characteristic of our culture. Mythic elements arise around historical and popular heroes as well: there are many myths associated with Abraham Lincoln and George Washington, for example.

Rather than being dismissible untruths, myths express some of the most powerful and important ideas in a society. In societies dependent on oral tradition rather than writing, myths reinforce values and ideals and help to transmit them from generation to generation. Myths in this sense are "true" even if they are fantastic and deal with impossible events, or have actors who could not have existed—like talking steam engines. Because myths encapsulate important cultural truths, anthropologists recognize that they are vitally important to a society and deserve respect. In the anthropological study of cultures, the term "myth" is not a pejorative. It is arguable that myths actually are more important than science!

Some myths are secular, others are religious, but all involve a symbolic representation of some societal or human truth. In the mythology of the ancient Greeks, the goddess Persephone joins her husband Hades below the surface of Earth for part of the year. When she is gone, her mother, Demeter, goddess of growing things, laments her absence, and winter comes. In the spring, when Persephone rejoins her mother, the world becomes green and fertile again. The story of Persephone and Hades not only symbolizes the passage of seasons; it also is a metaphor of the human realities of death and birth. Chinese culture reflects a strong sense of the importance of balancing opposites: yin/yang, light/dark, hot/cold, good/evil, wet/dry, earth/sky, female/male—there are many examples of this duality. A Chinese origin myth reflects this important cultural concern of balance: the creator god Phan Ku separates chaos into these opposites and establishes a series of dualities, including the separation of earth from sky, and other elements of the physical universe.

Some cultures have myths about creator figures or heroes who establish legitimacy for tribes or kin groups within a tribe by giving certain people particular lands, objects, or rituals that only they can use (Leeming and Leeming 1994). The telling of these myths may be incorporated in rituals that remind people of the relationships *among* people in society, as well as relationships between groups. So myths become a form of literature as well as a means to promote the continuity of a culture: stories are more meaningful and much easier to remember than lectures—a principle doubtlessly recognizable to anyone who has been a student!

Just as do tools and language, myths spread from people to people in a process anthropologists call diffusion. Flood stories are prominent in the Fertile Crescent area of the Middle East and into the Indus River area of India. Of course, humans must live near water, and after agriculture was invented, human settlements tended to congregate in river valleys, where control of water sources for agriculture often was the basis for political and religious power. Floods are not uncommon in such environments, and overflowing rivers may be a source of the fertility that attracts people to such settings. So it is not surprising to find that the early agricultural societies of the Middle East all possessed versions of a flood myth and a hero who survived it on a raft or boat: the Babylonians (Utnaptishtim), Sumerians (Ziusudra), Indians (Manu), Greeks (Deucalion and Pyrrha), and the Hebrews (Noah). Similarities in the flood myths of all of these groups suggest considerable diffusion—but there are differences as well, presumably reflecting individual cultural elements. After all, myths are symbolic of what is important to a people—and there were considerable differences between what was important to the Babylonians and what was important to the Hebrews, to take just one pair.

Sometimes as cultures come in contact with one another, new ideas and practices replace old ones, but more frequently cultural elements are borrowed and recombined. When Christian versions of creation from Genesis were encountered by the African Efe people, what eventually emerged was a combination origin myth incorporating a traditional female moon figure who helps the high god create human beings. He commands the people not to eat the fruit of the tahu tree, but one of the women disobeys. The moon sees her and reports her to the high god, who punishes human beings with death. If you are familiar with the biblical Adam and Eve story, you can see how elements were adapted by the Efe.

Types of Origin Myths

Although origin myths are quite varied, they can be grouped into types. The origin myth of the Colombian Cuebo Indians presents the world as always having existed, without a specific origin event, but most myths include a beginning time or event. Several cultures believe that in the beginning was a "cosmic egg," which either breaks like a familiar bird's egg to let forth a creator god (the Chinese Phan Ku, the Polynesian Taaroa, or the Hindu Prajapati) or is itself laid by a deity and hatches into elements of the universe. The Pelasgians of ancient Greece, for example, featured a cosmic egg laid by the goddess Eurynome, which hatched into the sun, moon, and stars as well as plants and animals (Leeming and Leeming 1994).

The beginning period might be a time of chaos, usually watery and dark, with supernatural beings emerging from a void. Perhaps reflecting a normal human preference for order and predictability over disorder and chaos, many origin myths attempt to explain how an orderly, understandable world emerged from frightening, formless disorder. Many traditions, such as that of the Hopi Indians, speak of a time when human beings lived underground and emerged to the upper world when led there by a spirit figure or god. Many origin myths describe the creation of Earth as resulting from the dismembering of a god or previous spirit: the Norse god Odin creates the mountains, seas, and other geographical features from the body of the slain giant Ymir; the Babylonian god Marduk creates the world from the body of the slain mother figure Tiamat.

The origin myths of North American Indian groups frequently include the "earth diver" motif, where a god or messenger is commanded to dive into the formless waters and bring up mud or silt which is made into dry land. Earth-diver myths are extraordinarily common, from Eastern Europe throughout Asia and into North America. The motif is even found in some Melanesian tribes of the Pacific.

Genesis Symbolism

The story of creation in the biblical book of Genesis symbolizes many things to people of Abrahamic faiths. Because they were migratory, and because they were located at a geographical crossroads, ancient Hebrews encountered many other Middle Eastern groups; as is typical in culture contact, they borrowed from neighbors as well as sharing their heritage. Origin myths of most of the Middle Eastern cultures, for example, included the motif of the creation of humans from clay. The primordial, chaotic state was one of water. The Genesis creation story derives in part from earlier Middle Eastern traditions from Babylonia and Persia, but with important differences.

According to the theologian Conrad Hyers, the ancient Hebrews found themselves surrounded by other tribes in which multiple gods were worshiped, a practice called polytheism. The Hebrews were also from time to time conquered by Egyptians, Assyrians, Babylonians, and Persians, which meant that remaining true to their traditions and avoiding absorption was a constant challenge. Of central importance to the Hebrews, and their major distinction among their neighbors, was their belief in one god (monotheism), and maintaining this belief (especially in the face of conquest) was difficult. There was much pressure upon the Hebrews to adopt the gods and idols of their neighbors. According to Hyers, the religious meaning of Genesis is largely to make a statement to both Hebrews and surrounding tribes that the one God of Abraham was superior to the false gods of their neighbors: sky gods (the sun, the moon, and stars), earth gods, nature gods, light and darkness, rivers, and animals (Hyers 1983). As Hyers puts it:

> Each day of creation takes on two principal categories of divinity in the pantheons of the day, and declares that these are not gods at all, but . . . creations of the one true

> God who is the only one. . . . Each day dismisses an additional cluster of deities, arranged in a cosmological and symmetrical order. (Hyers 1983: 101)

So on day 1 ("Let there be light"), God vanquishes the pagan gods of light and darkness. Similarly, gods of the sky and seas are displaced on day 2, while Earth gods and gods of vegetation are done away with on the third day. On the fourth day God creates the sun, moon, and stars, thereby establishing His superiority to them, and the fifth day removes divinity from the animal kingdom. Finally, on the sixth day God specially creates human beings, which takes away from the divinity of kings and pharaohs—but because God creates humans as his own special part of creation ("in His image"), *all* human beings are in some degree divine.

Genesis also described the nature of the Hebrew God. Unlike the gods of other Middle Eastern groups, the Hebrew God was ever present. Unlike the high god Marduk of the Mesopotamians, the Hebrew God did not originate from the actions of some other god or preexisting force. Genesis also makes it clear that God is omnipotent; unlike the Mesopotamian or Sumerian gods, the Hebrew God does not require preexisting materials from which to assemble Creation, but speaks (wills) the universe into being. God is also moral, being concerned with good and evil, contrasting strongly with the gods of the Hebrews' neighbors, who seem to govern in a universe that has little meaning or purpose. The Bible's God also is not part of nature, as some of the gods of others, but stands outside of nature as its creator (Sarna 1983).

Genesis also tells of the nature of humankind, "a God-like creature, uniquely endowed with dignity, honor, and infinite worth, into whose hands God has entrusted mastery over His creation" (Sarna 1983: 137). God forms the universe, making Earth the most important component and humans as its most important creature, having been given dominion over all other creatures and Earth itself. Humanity's responsibility is to husband Earth, but also to worship and obey God. Much of Genesis, especially the stories of Adam and Eve and of Noah and the Flood, reflect these themes; Adam and Eve are cast out of Paradise for disobeying God, and Noah is rewarded for his obedience and faith by being chosen to survive the Flood.

Thus Genesis reflects the character of a classic origin myth: it presents in symbolic form the values ancient Hebrews felt were most important: the nature of God, the nature of human beings, and the relationship of God to humankind. Hebrews distinguished their God from those of their neighbors and presented His deeds in their oral traditions and, eventually, in written form. Some of these writings became the Bible.

Modern Jews, Christians, and Muslims all revere the Bible as a sacred book, but each of the Abrahamic faiths has different interpretations of many of the events depicted—and differences of interpretation occur within the three faiths as well. For example, contrasting with the early Hebrew view, some modern Christians and Jews do not necessarily see God as separate from His Creation. There are also differences among sects as to the amount that God is believed to intervene in the world, and the nature and even the existence of miracles. Yet as did the ancient Hebrews, the Abrahamic faiths generally agree that God is omnipotent and good and that human beings are responsible to God. As will be discussed below, there are vast differences among believers as to specifics of faith, such as how literally the Bible should be read. Christians, Jews, and Muslims all have constituent sects that demand that the Bible be read literally, and all have sects that feel many or most passages should be read symbolically.

AMERICAN RELIGIONS

Americans practice a large number of religions, but the religion with the most adherents by far is Christianity. According to several polls, upward of 85 percent of Americans describe themselves as Christian. The largest survey of American religious views was conducted by scholars at the City University of New York in 1990. In the National Survey of Religious Identification (NSRI) researchers conducted a telephone survey of 113,723 adults, randomly chosen, with results statistically weighted to reflect American demographic characteristics (Kosmin and Lachman 1993). The percentage of error in a survey of this size is less than 0.5 percent.

Respondents were asked a simple question—"What is your religion?"—and answers, as well as information on geographic location, age, sex, income, and so on, were tabulated. The results of the survey are presented in Table 3.1 in the "1990" column.

The religious profile of Americans found in the 1990 NSRI study is reflected in other surveys conducted during the 1990s. A 1996 poll conducted by the humanist publication *Free Inquiry* found 90.7 percent of Americans stating they have a religion, with 83.8 percent claiming to be either Catholic or Protestant (Anonymous 1996). A Gallup poll conducted in December 2001 found 90 percent of Americans identifying themselves as having a "religious preference" and only 7 percent stating they had none (Gallup 2001).

However, a 2001 follow-up survey by the NSRI investigators showed changes in this religious profile. Using a smaller but still quite large sample of 50,281 individuals, investigators found that the percentage of Americans professing belief in God had declined, as had the percentage of Christians (Kosmin et al. 2002). The percentage of believers had declined from 89.5 percent to 80.2 percent, and the percentage of Christians had decreased from 86.2 percent to 76.5 percent. The largest increase was in the percentage of nonbelievers, which increased from 8.2 percent in 1990 to 14.1 percent in 2001. The "2001" column of Table 3.1 presents these more recent data.

Table 3.1
American Religious Profiles, 1990 and 2001

	1990 (%)	2001 (%)
Religious	89.5	80.2
Christian	86.2	76.5
Non-Christian	3.3	3.7
Jewish	1.8	1.3
Muslim	0.5	0.5
Other non-Christian	1.0	1.9
No Religion	8.2	14.1
Refused to State	2.3	5.4

Sources: 1990: Kosmin and Lachman 1993; Kosmin et al., 2001.

But whether the percentage of Christians is in the 80s or the 70s, it is nonetheless true that Christians are the largest religious group in the United States.

Christians can be further broken down into conservative or born-again Christians on the one hand, and mainstream Christians on the other. Conservative Christians are those who believe they have a *personal* relationship with Jesus, and who tie salvation to this belief. A higher percentage of conservative Christians than mainstream Christians regard the Bible as being literally true, according to a poll conducted by the Barna organization (Barna 2001). Most conservative Christians are Protestants, but some Catholics hold the same beliefs, especially those embracing charismatic Catholicism.

Antievolutionism in North America is rooted in religiously conservative Christianity; there are few if any activist Jews or Muslims opposing evolution in North America, and only small antievolution movements in Islamic countries such as Turkey and in the Jewish state of Israel. Although minority religions are growing in the United States, it is clear that Christianity is now, and for the near and intermediate future will be, the predominant American religious tradition. Because of their numbers and their prominence in the antievolution movement, the rest of this chapter will concentrate on Christians.

Many people are under the impression that there is a dichotomy between evolution and Christianity, a line in the sand between two incompatible belief systems. People with this belief feel that a person must choose one side of the line or the other. In reality, Christians hold many views about evolution, and Christian views actually range along a continuum, rather than being separated into a dichotomy.

THE CREATION/EVOLUTION CONTINUUM

Figure 3.1 presents a continuum of religious views with creationism at one end and evolution at the other. The most extreme views are, of course, at the ends of the continuum. The creation/evolution continuum reflects the degree to which the Bible is interpreted to be literally true; with the greatest degree of literalism at the top.

Figure 3.1
The Creation/Evolution Continuum. Courtesy of Alan Gishlick

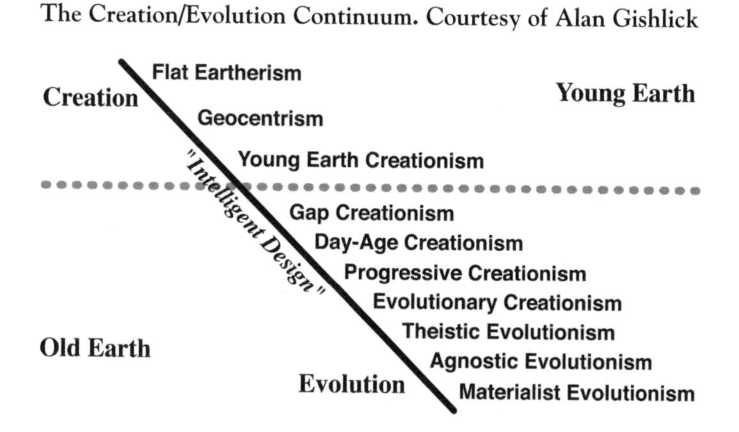

Although it is a continuum of religious and philosophical beliefs, it inversely reflects how much modern science is accepted by holders of these different views. I will begin with the strictest biblical literalists, the Flat Earthers. (For readers not familiar with the Bible, references take the form of book chapter: verse; thus, Genesis 1:4 would refer to the book of Genesis, chapter 1, verse 4.)

Flat Earthism

Until his death in March 2001, Charles K. Johnson of Lancaster, California, was the head of the International Flat Earth Research Society, an organization with a claimed membership of 3,500 (Martin 2001) which may not long outlive its leader's demise. Mr. Johnson—and we assume the members of his society—were very serious about their contention that the shape of Earth is flat, rather than spherical, because they are the most strict of biblical literalists. Few other biblical literalists hold to such stringent interpretations of the Bible. To Flat Earthers, many passages in the Bible imply that God created an Earth that is shaped like a coin, not a ball: flat, and round at the edges. Earth's disklike (not spherical) shape reflects biblical passages referring to the "circle" of the Earth (Isaiah 40: 22), and permits sailing around the planet and returning to one's starting point: one merely has to sail to the edge of Earth and make the circuit.

Because in this interpretation the Bible must be read as literally true, Earth must be flat (Schadewald 1991). The Englishman responsible for the nineteenth-century revival of Flat Earthism, Samuel Birley Rowbotham, "cited 76 scriptures in the last chapter of his monumental second edition of *Earth Not a Globe*" (Schadewald 1987: 27). Many of these refer to "ends of the Earth" (Deuteronomy, 28:64; 33:17; Psalms 98:3; 135:7; Jeremiah 25:33) or "quadrants" (Revelation 20:8). For Flat Earthers—and other literalists—the Bible takes primacy over the information provided by science; thus, because modern geology, physics, biology, and astronomy contradict a strict biblical interpretation, these sciences are held to be in error.

Geocentrism

Geocentrists accept that Earth is a sphere but deny that the sun is the center of the solar system. Like Flat Earthers, they reject virtually all of modern physics and astronomy as well as biology. Geocentrism is a somewhat larger, though still insignificant, component of modern antievolutionism. At the Bible-Science Association creationism conference in 1985, the plenary session debate was between two geocentrists and two heliocentrists (Bible-Science Association 1985). Similarly, as recently as 1985, the secretary of the still influential Creation Research Society was a published geocentrist (Kaufmann 1985).

Both Flat Earthers and geocentrists reflect to a greater or lesser degree the perception of Earth held by the ancient Hebrews, which was that it was a disk-shaped structure (Figure 3.2). They believed that the heavens were held up by a dome (*raqiya* or firmament) that arched over the land, and the land was surrounded by water. The firmament was perceived as a solid, metal-like structure which could be hammered and shaped (as in Job 37:18: "Can you beat out [*raqa*] the vault of the skies, as he does, hard as a mirror of cast metal?" [New English Bible]). The surface of the firmament is solid enough that God can walk on it (as in Job 22:14: "He walks to and fro on the dome of heaven" [New English Bible]). The sun, moon, and stars were attached to the firmament, which means that these heavenly bodies circled Earth *beneath* the firmament and, hence, were part of a geocentric universe. Further support for the idea of a solid sky and a geocentric solar system is found in Revelation 6:13–16: ". . . the stars in the sky fell to the earth, like figs shaken down by a gale; the sky vanished, as a scroll is rolled up. . . ." Stars were regarded as small, bright objects rather than massive suns hugely larger than Earth. They could fall upon Earth because they were below the firmament, a solid object which, if rolled aside, would reveal the throne of God (Schadewald 1987, 1981–1982).

The Bible also speaks of the "waters above the firmament"; ancient Hebrews conceived of the firmament supporting a body of water which came to earth as rain through the "windows of heaven" and was also the source of the 40 days and nights of rain of Noah's Flood.

Ancient and modern geocentricity reflects the idea that humans, other living things, and Earth are central to God, so to symbolize this importance God would make Earth the center of the universe. Taking Earth out of this central position reduces its importance, which reduces (according to some) man's place as the most important element in creation. Although not supporting geocentrism, the Institute for Creation

Figure 3.2
An early twentieth century conceptualization of ancient cosmology. (From Robinson, 1913 frontispiece)

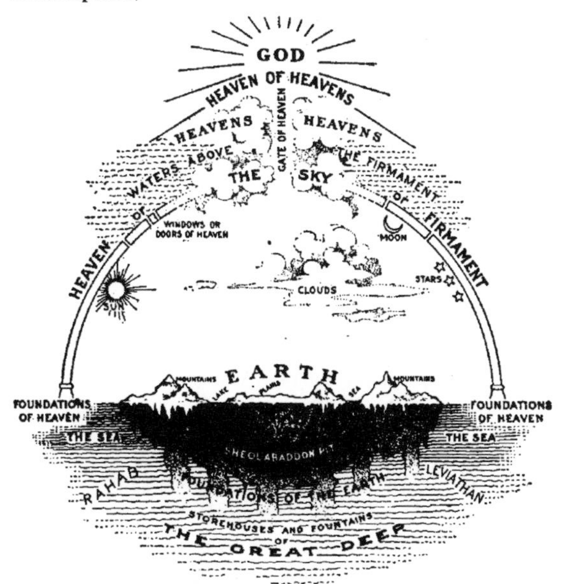

Research has promoted the same idea of the centrality of Earth and humans by claiming that our galaxy is actually central to the universe, even if our planet is not the center of our solar system (Humphreys 2002).

The next group of creationists on the continuum are less biblically literal than the previous two, but all three endorse a theological doctrine called Special Creationism. Special Creationism is the view that God created the universe, Earth, plants and animals, and humans during the course of six days, as indicated by a literal interpretation of Genesis. It is more than the conclusion that God created: it makes very specific claims about *how* God created. The most common form of Special Creationism holds that the creation event took place relatively recently, and is thus called Young Earth Creationism.

Young Earth Creationism

Few proponents of Young Earth Creationism (YEC) interpret the flat Earth and geocentric passages of the Bible literally. They accept heliocentrism, but they reject the results of modern physics, astronomy, chemistry, and geology concerning the age of Earth, and they deny biological descent with modification. Earth, in their view, is 6,000–10,000 years old. Although comparatively radical YECs will accept an age of Earth of 15,000 years, none accept the standard scientific view of billions of years of Earth history. They reject the big bang and postulate catastrophic mechanisms as the cause of most of the world's geological features. The Flood of Noah, for example, is supposedly responsible for carving the Grand Canyon and other geological features.

YECs reject the inference that earlier forms of life are ancestral to later ones: "kinds" are separate creations; descent with modification occurs only within "kinds," which are viewed as genetically limited. The definition of kinds is inconsistent but usually refers to a higher taxonomic level than species. Most YECs accept that God created creatures possessing at least as much genetic variation as occurs within a biological family (such as Felidae [the cat family] or Bovidae [the cattle family]), and then considerable "evolution within a kind" occurred. The created cat kind thus would have possessed sufficient genetic variability to differentiate into lions, tigers, leopards, pumas, bobcats, and house cats, through the normal microevolutionary processes of mutation and recombination, natural selection, genetic drift, and speciation. The basic body plans of major phyla that appear in the "Cambrian Explosion" are seen by most YECs as evidence of Special Creation.

The term Young Earth Creationist is often associated with the followers of Henry Morris, founder and recently retired director of the Institute for Creation Research (ICR) and arguably the most influential creationist of the second half of the twentieth century. He and John C. Whitcomb published *The Genesis Flood*, a seminal work that claimed to provide the scientific rationale for Young Earth Creationism (Whitcomb and Morris 1961). As the title suggests, the authors accept Genesis literally, including not just the special, separate creation of humans and all other "kinds," but also the historicity of Noah's Flood. Whitcomb and Morris proposed that there is scientific evidence to demonstrate the truth of Special Creationism: Earth is young, the universe appeared in essentially its present form about 10,000 years ago, and plants and animals appeared in their present forms as created kinds, rather than having evolved over millions of years through common ancestors. Although efforts to insist that a literal interpretation of the Bible is compatible with science, especially geology, occurred throughout the eighteenth and nineteenth centuries, *The Genesis Flood* was the first twentieth-century effort to attract a large following. Religious antievolutionists were greatly encouraged by the thought that there might be evidence that evolution was not only religiously objectionable but also scientifically flawed. Creation Science has been augmented by hundreds of books and pamphlets written by Morris and those inspired by him (McIver 1988).

The Institute for Creation Research (ICR) remains the flagship creationist institution to which all other YEC organizations look. It founded a publishing arm (Master Books) and maintains a graduate school offering master's degrees in science and science education, as well as a public museum. Most other YEC organizations sell and otherwise distribute ICR books, pamphlets, filmstrips, videos, movies, and other materials through their newsletters, and the movement leans heavily on Morris's writings and perspectives (Toumey 1994). The ICR also sponsors "Back to Genesis" and other revivals, hosted primarily in churches, during which ICR faculty lecture for one to three days. Thousands of people may attend these sessions, including many children from Christian schools and many home-schooled children. Other outreach activities include radio programs broadcast on several Christian radio networks, and occasional tours of the Grand Canyon and other sites. In proportion to the mission activity, little scientific research is performed by ICR faculty.

Old Earth Creationism

As mentioned, the idea that Earth is ancient was well established in science by the mid-1800s and was not considered a radical idea in either the Church of England or the Catholic Church (Eiseley 1961). From the mid-1700s on, the theology of Special Creationism has been partly harmonized with scientific data and theory showing that Earth is ancient. To many Christians, the most critical element of Special Creation is God's personal involvement in Creation; precise details of *how* God created are considered secondary. The present may indeed be different from the past, but Old Earth Creationists (OECs) see God as a direct causal agent of the observed changes.

The creation/evolution continuum, like most continua, has few sharp boundaries. Although there is a sharp division between YEC and OEC, the separation among the various OEC persuasions are less clear-cut. Even though OECs accept most of modern physics, chemistry, and geology, they are not very dissimilar to YECs in their rejection of biological evolution. There are several religious views that can be classed as OEC.

Gap Creationism. One of the better-known nineteenth-century accommodations allowing Christianity to accept the science of its time was Gap or Restitution Creationism, which claimed that there was a large temporal gap between verses 1 and 2 of chapter 1 of Genesis (Young 1982). Articulated from approximately the late eighteenth century on, Gap Creationism assumes a pre-Adamic creation that was destroyed before Genesis 1:2, when God re-created the world in six days and created Adam and Eve. A time gap between two separate creations allows for an accommodation of Special Creationism with the evidence for an ancient age of Earth. In Gap Creationism, the six days of Genesis 1:2 and following are considered to be 24-hour days.

Day-Age Creationism. Another attempt to accommodate science to a literal, or mostly literal, reading of the Bible is the Day-Age theory, which was more popular than Gap Creationism in the nineteenth century and the earlier part of the twentieth (Young 1982). Here religion is accommodated to science by having each of the six days of creation be not 24 hours but long periods of time—even thousands or millions of years. This allows for recognition of an ancient age of Earth but still retains a quite literal interpretation of Genesis. Many literalists have found comfort in what they interpret to be a rough parallel between organic evolution and Genesis, in which plants appear before animals, and human beings appear afterward. Anomalies such as flowering plants being created before animals, and birds occurring before land animals—incidents unsupported by the fossil record—are usually ignored.

Progressive Creationism. Although some modern activist antievolutionists may still hold to Day-Age and Gap views, the view held by the majority of today's Old Earth Creationists is some form of Progressive Creationism (PC). The PC view accepts more of modern science than do Day-Age and Gap Creationisms: Progressive Creationists do not dispute scientific data concerning the big bang, the age of Earth, or the long period of time it has taken for Earth to come to its current form. Indeed, some cite the big bang as confirmation of Genesis, in that the big bang is viewed as the origin of matter, energy, and time, which in the PC view is equivalent to *creation ex nihilo*, the doctrine of "creation out of nothing." As in other forms of OEC, although theories of modern physical science are accepted, only parts of modern biological science are incorporated into PC.

For example, the fossil record shows a consistent distribution of plants and animals through time: mammals are never found in the Cambrian, for example, and flowering plants are never found in the Devonian. YECs believe that flowering plants, dinosaurs, humans, and trilobites were all created at the same time and therefore all lived at the same time. They regard the orderly distribution of fossils in strata around the world to be an artifact of Noah's Flood, which is thought to have differentially sorted organisms into groups, even if they all died at the same time. PCs, on the other hand, generally accept the fossil distribution of organisms as "real" because they believe God created "kinds" of animals sequentially. To PCs, the geological column reflects history: God first created simple, single-celled organisms, then more complex ones, then simple multicellular organisms, then more complex ones, and so on up until the present time. With PC, there is no difficulty with seed-bearing plants appearing after ferns and cycads: God created the more "advanced" plants at a later time. PCs do not, however, accept that the "kinds" evolved from one another, though they are no more specific about what constitutes a "kind" than are YECs. As in YEC, though, a "kind" is viewed as being genetically limited: as a result, one kind *cannot* change into another.

Evolutionary Creationism. Despite its name, evolutionary creationism (EC) is actually a type of evolution. Here, God the Creator uses evolution to bring about the universe according to His plan. From a scientific point of view, evolutionary creationism is hardly distinguishable from Theistic Evolution, which follows it on the continuum. The differences between EC and Theistic Evolution lie not in science but in theology, with EC being held by more conservative (Evangelical) Christians, who view God as being more actively involved in evolution than do most Theistic Evolutionists (D. Lamoureux, personal communication).

Intelligent Design Creationism has been positioned on the continuum as overlapping YEC and OEC because some of its proponents can be found in each camp.

Intelligent Design Creationism

Intelligent Design Creationism (IDC) is the newest form of creationism, and yet it resembles a much earlier idea. In some ways, IDC is a descendant of William Paley's Argument from Design (Paley 1803), which argued that God's existence could be proved by examining His works. Paley used a metaphor: if one found a watch, it was obvious that such a complex object could not have come together by chance; the existence of a watch implied a watchmaker who had designed the watch with a purpose in mind. By analogy, the finding of order, purpose, and design in the world was proof of an omniscient designer.

The vertebrate eye was Paley's classic example of design in nature, well known to educated people of the nineteenth century. Because of its familiarity, Darwin deliberately used the vertebrate eye in *On the Origin of Species* to demonstrate how complexity and intricate design *could* come about through natural selection and did not require divine intervention.

> To suppose that the eye, with all its inimitable contrivances for adjusting the focus to different distances, for admitting different amounts of light, and for the correction of spherical and chromatic aberration, could have been formed by natural selection, seems, I freely confess, absurd in the highest possible degree. Yet reason tells me, that if numerous gradations from a perfect and complex eye to one very imperfect and simple, each grade being useful to its possessor, can be shown to exist; if further, the eye does vary ever so slightly, and the variations be inherited, which is certainly the case; and if any variation or modification in the organ be ever useful to an animal under changing conditions of life, then the difficulty of believing that a perfect and complex eye could be formed by natural selection, though insuperable by our imagination, can hardly be considered real. (Darwin 1966: 186)

Structures and organs that accomplish a purpose for the organism—allow capture of prey, escape from predators, or attracting a mate—could be designed directly by an omniscient designer, or they could be "designed" by a natural process that produced the same effect. As will be discussed in more detail elsewhere in this book, Darwin's argument that a natural process such as natural selection could explain apparent design was theologically offensive to those who believed that God created *directly*.

In IDC one is less likely to find references to the vertebrate eye and more likely to find molecular phenomena such as DNA structure or cellular mechanisms held up as too complex to have evolved "by chance." The IDC high school biology supplemental textbook, *Of Pandas and People* (Davis and Kenyon 1993), weaves allusions to information theory into an exposition of the "linguistics" of the DNA code in an attempt to prove that DNA is too complex to explain using natural causes.

In the PC tradition, IDC proponents accept natural selection but deny that mutation and natural selection are adequate to explain the evolution of one kind to another, such as chordates from echinoderms or humans from apes. The emergence of major anatomical body types and the origin of life, to choose just two examples popular among IDCs, supposedly are phenomena "too complex" to be explained naturally; thus IDC demands a role be left for the intelligent designer—God. IDC will be discussed in more detail in chapter 7.

Theistic Evolutionism

Theistic Evolution (TE) is a theological view in which God creates through the laws of nature. TEs accept all the results of modern science, in anthropology and biology as well as astronomy, physics, and geology. In particular, it is acceptable to TEs that one species can give rise to another; they accept descent with modification. TEs vary in whether and how much God is allowed to intervene—some believe God created the laws of nature and is allowing events to occur with no further intervention. Other TEs see God as intervening at critical intervals during the history of life (especially in the origin of humans). In a recent book, an entire continuum of Theistic Evolutionists is presented; clearly, there is much variation (Hewlett and Peters 2003). In one form or another, TE is the view of creation taught at the majority of mainline Protestant seminaries, and it is the position of the Catholic Church. In 1996, Pope John Paul II reiterated the Catholic version of the TE position, in which God created, evolution happened, humans may indeed be descended from more primitive forms, but the Hand of God was required for the production of the human soul (John Paul II 1996).

In Figure 3.1, TE is followed by two nontheistic views, Agnostic Evolutionism (AE) and Materialist Evolutionism (ME).

Agnostic Evolutionism

Although poll data indicate that most Americans have a belief in God or some higher power, a minority do not (Kosmin et al. 2002). Just as there are variations in worldview among believers, so also are there differences among those who do not believe in God. The term "agnostic" was coined by "Darwin's Bulldog," the nineteenth-century scientist Thomas Henry Huxley, to refer to someone who suspended judgment about the existence of God. Huxley felt that in this world, it is impossible to know or even grasp ultimate reality; therefore neither belief in nor rejection of the existence of God is warranted. To Huxley, the thoughtful person should suspend judgment. Huxley was a strong supporter of science and believed that knowledge and beliefs should be based upon empirical knowledge—and that supernaturalism would eventually be supplanted by science. But he felt it was more honest not to categorically reject an ultimate force or power beyond the material world (Huxley [2002] 1884).

> I have no doubt that scientific criticism will prove destructive to the forms of supernaturalism which enter into the constitution of existing religions. On trial of any so-called miracle the verdict of science is "Not proven." But true Agnosticism will not forget that existence, motion, and law-abiding operation in nature are more stupendous miracles than any recounted by the mythologies, and that there may be things, not only in the heavens and earth, but beyond the intelligible universe, which "are not dreamt of in our philosophy." The theological "gnosis" would have us believe that the world is a conjuror's house; the anti-theological "gnosis" talks as if it were a "dirt-pie" made by the two blind children, Law and Force. Agnosticism simply says that we know nothing of what may be beyond phenomena.

Agnostics believe that in this life, it is impossible truly to know whether there is a God, and although they believe that it is not probable that God exists, they tend not to be dogmatic about this conclusion. AEs accept the scientific evidence that evolution occurred, but they do not consider important the question of whether God is or was or will be involved. They differ from the next position on the continuum by not categorically ruling out the involvement of God, although like Materialist Evolutionists, they are nonbelievers.

Materialist Evolutionism

We should distinguish between two uses of the term "materialism" (or "naturalism"). As we discussed earlier, modern science operates under a rule of *methodological naturalism* that limits it to attempting to explain natural phenomena using natural causes. Materialist Evolutionists (ME) go beyond the methodological naturalism of science to propose not only that natural causes are sufficient to explain natural phenomena, but also that the supernatural does not exist. This is a form of *philosophical naturalism*. To a philosophical naturalist, there is no God. The philosophy of humanism is a materialistic philosophy, as is atheism. As discussed earlier in this chapter, philosophical naturalism is distinct from the practical rules of how to do science.

This is an important distinction to the subject of this book because some anti-evolutionists criticize evolution and science in general for being not only methodologically naturalistic but also philosophically naturalistic. This is a logical error, as can be seen in Figure 3.3. It is very likely the case that all philosophical naturalists are simultaneously methodological naturalists (all Ps are Ms). It does not follow that all methodological naturalists are philosophical naturalists (not all Ms are Ps). It *might* be the case—if both circles were the same size and right on top of one another—but this would have to be determined empirically, not logically. In fact, such a claim is empirically falsified, for there are many scientists who accept methodological naturalism in their work, but who are theists and therefore not philosophical naturalists. Gregor Mendel—the monk whose research became the foundation of genetics—is a classic case of a scientist who was a methodological naturalist but not a philosophical one, and there are many scientists today who, like him, are methodological but not philosophical naturalists.

As mentioned, there are varieties of belief within the various positions on the continuum, and this is true for materialists as well. For example, although materialists

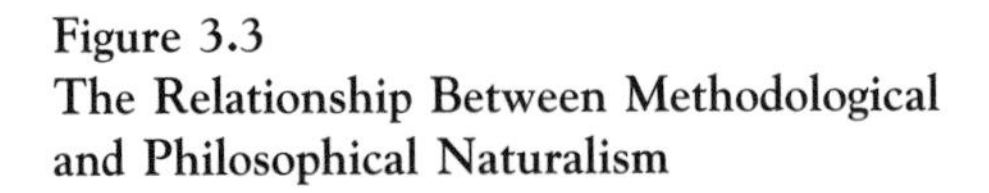

Figure 3.3
The Relationship Between Methodological and Philosophical Naturalism

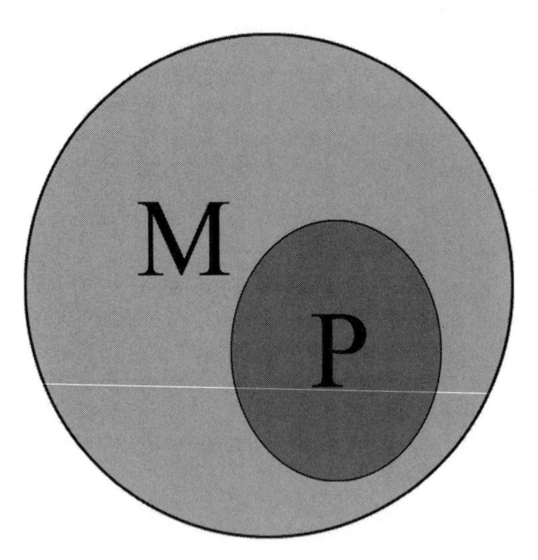

All philosophical naturalists are methodological naturalists, but it is not accurate to say that all methodological naturalists are philosophical naturalists. One can thus be a scientist practicing methodological naturalism but still be a theist.

share a high opinion of science and accept evolution, they do not all share the same attitudes toward religion. *Agnostic* materialists don't consider that the question of whether God created can be answered. *Humanists* believe that "Humanism is a progressive lifestance that, without supernaturalism, affirms our ability and responsibility to lead meaningful, ethical lives capable of adding to the greater good of humanity" (American Humanist Association 2002). The two major humanist organizations are the American Humanist Association, at the time of this writing consisting of approximately 5,000 members, and the Council for Secular Humanism, with approximately 4,000 members.

The third major group within materialists, *atheists*, reject the existence of God but tend to be more actively antireligious than the other two. There are about 2,200 members of the best-known atheist group, the American Atheists. Clearly, any single theist organization has far more members than all the materialist organizations combined. If nonbelievers are between 10 percent and 14 percent of the population, as suggested by some polls, the vast majority of them do not join groups of like-minded individuals.

This presentation of Christian and materialist views regarding creation and evolution is simplified—just as was the earlier presentation of the nature of science in chapter 1 and the presentation of the science of evolution in chapter 2. It is possible to go into far more detail on any of these beliefs, but a shorthand version will have to suffice to introduce the topic.

RELIGION, SCIENCE, AND PHILOSOPHICAL NATURALISM

What are the relationships among religion, science, and philosophical naturalism? Everyone recognizes that there are differences, but there are similarities as well. All three of these terms refer to ways of knowing: a field of study that philosophers call "epistemology." The epistemology we call science is primarily a methodology that attempts to explain the natural world using natural causes. Although individual scientists may be concerned with moral and ethical issues, or rules of conduct, science as a way of knowing is not. The methodology of testing natural explanations against the natural world will not tell us whether it is immoral for coyotes to kill rabbits, or whether members of one sex or another should keep their heads covered in public, or whether marrying your father's brother's child is immoral but marrying your father's sister's child is not. Science is actually a quite limited way of knowing, with limited goals and a limited set of tools to use to accomplish those goals.

Philosophical naturalism relies upon science and is inspired by science, but it differs from science in being concerned with rules of conduct, ethics, and morals. When a scientist makes a statement like "Man is the result of a purposeless and natural process that did not have him in mind" (Simpson 1967: 344), it is clear that he or she is speaking from the perspective of philosophical naturalism rather than from the methodology of science itself. As anthropologist Matt Cartmill has observed, "Many scientists are atheists or agnostics who want to believe that the natural world they study is all there is, and being only human, they try to persuade themselves that science gives them grounds for that belief. It's an honorable belief, but it isn't a research finding" (Cartmill 1998: 83). Philosophical naturalism is embraced by only a minority of Americans—perhaps as few as 10–14 percent or so—but it has had a long history in Western culture, going back to some of the pre-Socratic philosophers of ancient Greece.

Religion concerns the relationship of people with the divine, but it also may include explanations of the natural world and the origin of natural phenomena. Religious views almost universally derive from revelation, but this does not rule out the use of empirical and logical approaches to theology. In fact, many Christian denominations pride themselves on their reliance on logic and reason as a means both to understand the natural world and to evaluate theological positions. But an ultimate reliance on revelation can place religion into conflict with science, as discussed earlier in this chapter. When revealed truth conflicts with empirical knowledge, how does one choose?

Different religious traditions provide different interpretations of revealed truth—all held with equal fervor—and within the same religious tradition the documents which are considered authoritative can be, and usually are, interpreted differently by different adherents. Reform and Hasidic Jews interpret the Torah differently, Muslims of the Shiite and Sunni traditions have some different interpretations of the Koran, and Catholics and Protestants use Bibles with different books. Which tradition is more faithful to the sacred documents is ascertained differently by different factions, and unless agreement can be reached on criteria of judgment, different factions will be unable to determine whose interpretation is correct.

For example, some Christians interpret the Bible as indicating that the Flood of Noah was an actual historical event, covering the entire Earth, and believe that the Grand Canyon was cut by the receding Flood waters. Other Christians interpret the Bible differently and argue that the Flood was not a universal historical event and could not have carved the Grand Canyon. Proponents of different biblical interpretations tend to not persuade one another because their religious assumptions are different; to some it is not a matter of logic or empirical evidence (as will be illustrated in the readings in part III).

In science, on the other hand, there is no revealed truth. Although some explanations are believed to be very solidly grounded, it is understood that even well supported theories can be modified and, in rare circumstances, might even be replaced by other explanations. For the limited purpose of explaining the natural world, science has a major advantage over religion in that individuals of different philosophical, religious, cultural, and/or ideological orientations, using the methodology of science, can debate their differences based on repeatable—and repeated—empirical investigations. Different scientists, using different techniques, technologies, and observational approaches, provide validation not possible through revelation.

Scientists looking at geological and biological data can piece together a natural history of the Grand Canyon and test one another's explanations against the lay of the land itself. The ability to go back to nature—again and again—to test explanations, rework them, and retest them is one of the strengths of science and a major contributor to the amount of empirical knowledge exponentially amassed over the last 300 years. To some, though, the open-endedness of science is a weakness: they seek definite answers that will never change. For them, Ashley Montagu's definition of science as "truth without certainty" is insufficient; for others, it is science's greatest strength (Montagu 1984: 9).

Just as attempts to explain the natural world through revelation cause friction with scientists, so also do attempts by some materialist scientists—speaking in the name of science—who make statements about the ultimate nature of reality cause friction with religious people. Upon reflection it should be recognizable that if science has the limited goal of explaining the natural world using natural causes, it lacks the tools to make justifiable statements about whether there is or is not a universe beyond the familiar one of matter and energy. As will be clear in some of the readings to follow, both supporters and deniers of evolution argue erroneously that because science utilizes methodological naturalism (and quite successfully), science therefore also incorporates philosophical naturalism. Unfortunately, such a confusion makes communication about science and religion, or creationism and evolution, more difficult.

REFERENCES CITED

American Humanist Association. 2002. *Definitions of Humanism*. Accessed August 10, 2002. Available from http://www.americanhumanistorg/humanism/definitions.htm.

Anonymous. 1996. Religious Belief in America: A New Poll. *Free Inquiry*, Summer: 34–40.

Augustine, Saint. 1982. *The Literal Meaning of Genesis (De Genesi ad Litteram)*, translated by J. H. Taylor. Vol. 1. New York: Newman Press.

Barna, George. 2001. *Born-Again Christians*. Barna Research Group, April 19, 2001. Accessed October 19, 2003. Available from http://www.barna.org/cgi-bin/PageCategory.asp?CategoryID=8.

Bible-Science Association. 1985. Bible-Science Association Conference Schedule. Cleveland, OH: Bible-Science Association.

Carneiro, Robert. 1983. Origin Myths. Pamphlet. Berkeley, CA: National Center for Science Education, Inc.

Cartmill, Matt. 1998. Oppressed by Evolution. *Discover*, March: 78–83.

Darwin, Charles. 1966. *On the Origin of Species: A Facsimile of the First Edition*. Cambridge, MA: Harvard University Press.

Davis, Percival W., and Dean H. Kenyon. 1993. *Of Pandas and People*, 2nd ed. Dallas, TX: Haughton.

Eiseley, Loren. 1961. *Darwin's Century*. Garden City, NY: Doubleday.

Gallup, George. 2001. *Poll Topics and Trends*. Accessed January 1, 2002. Available from www.gallup.com/poll/topics/religion2.asp.

Hewlett, Martinez, and Ted Peters. 2003. *Evolution from Creation to New Creation*. Nashville, TN: Abingdon Press.

Humphreys, D. Russell. 2002. The Battle for the Cosmic Center. *Acts and Facts Impact*, August: a–c.

Huxley, Thomas H. 1884. Agnosticism: A Symposium. *The Agnostic Annual* 1. Also available on the Web at http://aleph0.clarku.edu/huxley/UnColl/Rdetc/AgnAnn.html.

Hyers, Conrad. 1983. Biblical Literalism: Constricting the Cosmic Dance. In *Is God a Creationist? The Religious Case Against Creation-Science*, edited by R. M. Frye. New York: Scribner's.

John Paul II. 1996. Magisterium. *L'Osservatore Romano*, October 30, 3, 7.

Kaufmann, D. A. 1985. Geocentricity: A Creationist Consideration. *The Christian News* 23 (21): 7.

Kosmin, Barry A., and Seymour P. Lachman. 1993. *One Nation Under God*. New York: Crown.

Kosmin, Barry A., Egon Mayer, and Arieta Keysar. 2002. *American Religious Identification Survey 2001* (pdf). Graduate Center of the City University of New York, 2001. Accessed March 10, 2002. http://www.gc.cuny.edu/studies/aris.pdf.

Leeming, David A., and Margaret Leeming. 1994. *A Dictionary of Creation Myths*. New York: Oxford University Press.

Martin, Douglas. 2001. Charles Johnson, 76, Proponent of Flat Earth. *New York Times*, March 25, Section 1, p.44.

McIver, Thomas. 1988. *Anti-Evolution: An Annotated Bibliography*. Jefferson, NC: McFarland.

Montagu, M. F. Ashley. 1984. *Science and Creationism*. New York: Oxford University Press.

Paley, William. 1803. *Natural Theology: Or, Evidences of the Existence and Attributes of the Deity, Collected from the Appearances of Nature*, 5th ed. London: Faulder.

Raman, V. V. 1998–1999. A Hindu View on Science and Spirituality. *Science and Spirit* 9 (5): 6–7.

Robinson, G. L. 1913. *Leaders of Israel*. New York: Association Press.

Sarna, Nahum M. 1983. Understanding Creation in Genesis. In *Is God a Creationist? The Religious Case Against Creation-Science*, edited by R. M. Frye. New York: Scribner's.

Schadewald, Robert J. 1981–1982. Scientific Creationism, Geocentricity, and the Flat Earth. *Skeptical Inquirer* 6 (2): 41–48.

Schadewald, Robert J. 1987. The Flat-Earth Bible. *Bulletin of the Tychonian Society* 44, July: 27–39.

Schadewald, Robert J. 1991. Introduction. In *A Reparation*, edited by C. S. De Ford. Washington: Ye Galleon.

Simpson, George Gaylord. 1967. *The Meaning of Evolution*, rev. ed. New Haven, CT: Yale University Press.

Stevens, P., Jr. 1996. Religion. In *Encyclopedia of Cultural Anthropology*. New York: Holt.

Toumey, Christopher. 1994. *God's Own Scientists*. New Brunswick, NJ: Rutgers University Press.

Whitcomb, John C., and Henry R. Morris. 1961. *The Genesis Flood: The Biblical Record and Its Scientific Implications*. Phillipsburg, NJ: Presbyterian and Reformed Publishing Company.

Young, Davis A. 1982. *Christianity and the Age of the Earth*. Grand Rapids, MI: Zondervan.

Part II

A History of the Creationism/ Evolution Controversy

The history of—and potential conflict between—creationism and evolution extends back hundreds of years. The current controversy has its roots in conflicting ideas of stasis and change that reach back beyond the Middle Ages. Darwin is unquestionably a central figure in the development of ideas of biological evolution, but he of course built on ideas of an ancient and changing physical universe and world presented by astronomers and geologists working as much as a century before him.

Part II provides an introduction to this history, beginning in chapter 4 with pre-Darwinian ideas about evolution and continuing with Darwin's ideas and their reception. Chapter 5 picks up at the beginning of the twentieth century with the antievolution movement that culminated in the Scopes trial, which was followed by a long period in which evolution was largely ignored in the public schools. With evolution's return to textbooks and the classroom in the late 1960s, we encounter the period of "creation science." Finally, chapter 6 describes the current "neocreationist" movement, which employs a set of antievolutionist strategies designed to avoid legal decisions that hamstrung the earlier creation science movement.

CHAPTER 4

Before Darwin to the Twentieth Century

BEFORE DARWIN: STASIS VERSUS CHANGE

Many tend to think of the creationism/evolution controversy as one of "God did it" versus "Natural processes did it," but that is a false dichotomy. As discussed in chapter 3, many religious people conceive of creation as the result of God working through natural processes. Historically, a more accurate distinction between creationism and evolution focuses on "what happened" rather than "who done it." Special Creationists view nature as largely static after the events of the Creation, whereas those who accept evolution view the universe as having a history: the present universe, the planet Earth, and the living things on it are different from the universe, Earth, and life of the past: change through time has taken place. However, the two models of stasis and change are not relegated to the current creationism/evolution controversy: they have deep historical roots.

Nature Through Time

Western concepts of nature and time have been shaped by the ideas of the philosophers Plato and Aristotle, as well as those of many Christian thinkers—even though sometimes these ideas were not in complete harmony. From Plato came the idea of *idealism*, the concept that the world and its objects as perceived by our senses were not "real," but only an imperfect copy of what existed in a "transcendent world of pure and immutable forms" (Durant 1998: 269). In *The Republic*, Plato uses the metaphor of our images of the world being similar to flickering shadows cast on a wall, unreal in themselves, with the true reality being the light which produces the shadows. Christian scholars reformulated idealism in terms of the Creation: God created the world according to a plan: there was an ideal form that lodged in the mind of God, and what we see in the real world are merely variants—imperfect copies in some cases—of that ideal. Dwelling as an idea in the mind of God is the ideal rabbit, human being, or barnacle: the variation in size, shape, color, and so on that we see in nature is less important than the true essence of rabbits, humans, and barnacles which resides in the mind of God. (A linguistic fossil of this view—though not carrying the same meaning—is the biological term "type specimen.")

From Aristotle came a view of nature that focused not only on form but also on function. Aristotle wrote of the *purpose* of nature: why something existed, not just what form it took. The rain falls *to make the grass grow*. Deer have long legs *to run away from predators*. Christian theology was also influenced by these ideas: humans exist because *they had been created by God to worship Him*. Explaining something by its purpose is known as *teleology*. It is understandable that form is related to purpose: form follows (and contributes to) function, after all. To allow it to escape from predators, the deer has long legs: the legs of a deer were *designed* to enable it to survive, as its teeth were designed to allow it to eat woody shoots—as the teeth of the wolf were designed to eat meat. Thus purpose (teleology) and design were linked concepts.

Reflecting his view of immutable forms, Aristotle classified plants and animals in terms of "kinds" of organisms that could be ranked in a linear "Great Chain of Being," or "Scale of Nature" (*scala naturae*). This essentialist view fit very comfortably with the Christian doctrine of Special Creationism. God created all creatures great and small, and simple to complex, and the span of created beings could be ranked hierarchically. Humans were almost at the top of the Great Chain of Being, right beneath angels, which in turn were second to God Himself at the very top of the hierarchy.

The doctrine of Special Creationism incorporated these Greek ideas—the hierarchical ordering of nature, and of design and purpose—and included the Christian idea of an omnipotent, omniscient Creator who stood outside of nature. In the theology of Special Creationism, God created the universe at one time (taking six days in the most common view, although, as discussed in chapter 3, Gap Creationism considered two special creations) in essentially its present form. He created living things as we see them today for a particular environment and way of life. He also created stars and galaxies as we see them, and created the planet Earth as we see it today, as the home of human beings and the creatures over which we have been given dominion and stewardship.

For most of European history, educated people blended the Christian and Greek views and concluded that the world was stable and largely unchanging. In such a conception, the age of Earth was unimportant: it was not until theologians untangling the genealogies of the Bible calculated that Earth was approximately 6,000 years old that anyone considered the question of the age of Earth even worth asking: the specially created, essentialist universe of stars, planets, Earth, and its inhabitants had come into being in its present form, and was assumed to be virtually the same as it had been at the Creation. The notion that Earth—much less living things—could have had a history was not entertained through the Middle Ages. Stasis extended even to medieval and feudal social life: everyone's place in society was determined by birth. Serfs were to serve, the nobility were to rule them, and kings had a divine right—God-given—to rule. The sociopolitical stasis of society mirrored the conception of an unchanging natural world, all created the way it was by God, for His purposes, whatever they may be.

But there was growing evidence that things might not be static after all—both socially and in the natural world. By the Renaissance in the 1500s, a middle class began growing and society looked rather less static (though Shakespeare's *Henry V* still reflects enthusiasm for the old view of the divine right of monarchs to rule). The expansion of societal boundaries continued through the Enlightenment of the 1700s, as cities grew, the merchant class expanded, and democratic ideas began to replace those of the divine right of the church and hereditary monarchs to rule.

The conception of nature as stable—and known—was troubled by the European discovery and exploration of North and South America and Oceania from about 1500 to 1800. The Age of Exploration exposed Europeans to huge unknown natural areas. Even if Columbus died thinking he had discovered a route to the Orient, it soon became clear to others that the animals, plants, people, and geological features he had encountered were truly from a New World. During the 1700s and through the 1800s, the study of nature—natural history—was a popular pastime of not only educated individuals but also ordinary citizens. The Swedish natural historian Carl von Linné (whose name is Latinized as Linnaeus) developed a useful classification system for plants and animals that grouped them into gradually broader categories: species were grouped into genera, genera into families, families into orders, orders into classes, and so on. He received specimens to classify from all over the world, sent not only by captains of exploring ships but even by traders and common seamen. Another "new world" became apparent with the invention of the microscope in the early 1600s and the discovery of the world of microorganisms. Europeans of the Enlightenment experienced an expansion of knowledge of the natural world that disrupted old ways of thinking, much as new economic and political systems disrupted the social systems of the day.

The discoveries of natural history had implications for Christian religious beliefs. Europe, Africa, and Asia were mentioned in the Bible, but the New World was not; thus the Bible did not contain all knowledge. Puzzles appeared: there were animal and plant species in North America and other new lands that were not found in the Old World, such as opossums, llamas, tobacco, tomatoes, potatoes, and corn. Had the newly discovered species been created at the same time as known ones? Had they merely died out in some places? In the early 1800s, the French comparative anatomist Georges Cuvier had determined that fossil bones found in Europe were indeed sufficiently similar to living forms to be classified as mammals or reptiles, and even more narrowly as elephants and other known varieties. Yet these bones were sufficiently different that it was clear that they came from species that no longer existed. The disappearance of huge reptiles (dinosaurs) and certain mammals, such as mammoths and saber-toothed cats, was unexplained. The notion that some "kinds" had become extinct was theologically troubling because of the implication that the creation might not have been perfect, which in turn generated problems for the concept of the original sin of Adam and Eve. Perhaps the species represented by the European fossils were actually still living in the New World—that would solve some theological problems. One of the instructions Thomas Jefferson gave to the explorers Meriweather Lewis and William Clark, in fact, was to keep watch for mammoths and other animals known only from the fossil record as they explored the western reaches of the North American continent. Cuvier himself argued that extinctions of some species had occurred and were the result of a series of environmental catastrophes. To some of the scientists of the day, the most recent of these catastrophes was Noah's Flood.

Even more difficult to explain—and creating theological problems in their own right—were the human inhabitants of the new lands. Native Americans, Polynesians, and other peoples new to Europeans were not mentioned in the Bible. Wild tales were told of one-eyed races, of people who barked like dogs or who were part animal, and other monstrous creatures. But real, undeniable human beings were encountered as well. How could they be explained? Were they also the children of Adam? Or were they creatures of Satan? Were they human? Did they have souls? Could they become Christian? In 1537, Pope Paul III declared that the Indians of the New World were indeed human and not animals—and therefore should not be enslaved (Gossett 1965: 13). They thus had souls and were fit subjects for Christianization. But how did they come to be living where they were found? If the Ark had landed at Ararat, how did the Indians get to the New World?

In 1665, Isaac de la Peyrère produced the first version of Gap Creationism (see chapter 3), proposing an explanation for these newly discovered peoples that was compatible with the Bible. He proposed that Genesis records two creations, the first being described in Genesis 1, and the second—the Adam and Eve creation—in Genesis 2. Native Americans, Polynesians, Australian Aborigines, and anyone else not specifically mentioned in the Bible were descendants of the first, or "pre-Adamite," creation. The pre-Adamites were also the source of Cain's wife—solving another theological problem. In the second, Adamic creation, Genesis 2 and following, God created anew, and Adam and Eve were the progenitors of the more familiar human beings in Europe, Asia, and Africa. Unfortunately, this theological view generated problems of its own, raising the issue of whether pre-Adamites were innocent of original sin. Presumably so—since they were unrelated to Adam—but then, were they in need of salvation by Jesus? The discovery of the New World required rethinking many Christian doctrines, as new facts had to be fit into old frameworks.

More new facts were forthcoming from the study of the Earth in the late 1700s. In Great Britain, William Smith was given the task of surveying the countryside preparatory to the excavation of a canal system across England (Winchester 2001). It was clear that Britain consisted of a variety of types of geological formations, some of which held water better than others, and it behooved the young surveyor to be able to identify and classify the various layers to ensure that canals functioned properly. He did a superb job, tracing strata for sometimes hundreds of miles across the countryside and making detailed maps. He made a discovery (confirmed by Cuvier and French geologists): different strata consistently contained different fossils, and he could in fact classify a stratum if he knew what kinds of fossils it contained, regardless of where it was found. He also showed that the deeper the layer, the more different the fossils were from living plants and animals. Many fossils—especially the deeper ones—were no longer represented by living animals. It seemed logical that, by and large, bottom layers were older than top layers; thus there were older animals that differed from more recent ones, and extinct animals that lived long ago. Estimates could be made of the length of time it took for a valley to erode or for a chain of mountains to lift up. Through careful description and logic, Smith demonstrated the principle that time and change are reflected in the rocks (Winchester 2001).

An appreciation also grew for the nature of geological processes such as sedimentation and erosion; the understanding that nature was dynamic rather than static began to grow as knowledge of the natural world—from geology as well as biology—increased through the 1700s and 1800s. Arguably, the view of nature as dynamic required the amassing of a critical amount of accurate information about the natural world which hadn't accumulated until the early 1800s. A relationship between geology, biology, and time began to be appreciated: by the mid-nineteenth century, Darwin's time, the once radical idea that Earth was really quite old, and had changed through time, was becoming well accepted in the scientific community and by educated people in general—including the clergy. If Earth had changed, could not other aspects of nature also have changed? Darwin's contribution to the growing appreciation that nature was dynamic rather than static was to add living things to the list of natural phenomena that changed through time.

WHAT DARWIN WROUGHT

Charles Darwin was a respected scholar and scientist well before the 1859 publication of his best-known book, *On the Origin of Species*. He had made his reputation first as a geologist, by providing a plausible (and correct) hypothesis about the formation of coral reefs. He then wrote about other geological topics such as volcanoes before turning his hand to biology. Darwin was a meticulous observer of nature (as seen in his four-volume study of the anatomy and physiology of barnacles, and his research on orchids) but also an experimentalist: at his country estate he had not only a small laboratory but also sufficient land to conduct experiments on varieties of plants. He maintained a voluminous correspondence with scientists of his day, and because he was so meticulous in his record-keeping, much of it remains for scholars to study (Burkhardt and Smith 2002).

On the Origin of Species was Darwin's ninth book of an eventual total of 19 books and monographs. The first printing of the *Origin* sold out rapidly, being bought not only by scientists but also by educated laymen and clergy. It sold steadily over the years, allowing Darwin to make corrections and small modifications in subsequent editions. There were six editions in all.

The Scientific Response to *On the Origin of Species*

Darwin made two major points in the *Origin*: that living things had descended with modification from common ancestors, and that the main mechanism resulting in evolution was the mechanism he had discovered, which he called "natural selection" (see chapter 2). As described by the historian Ronald Numbers (1992, 1998), in the late nineteenth and early twentieth centuries, scientists in the United States responded largely positively to Darwin's ideas. The idea of evolution itself was less controversial than Darwin's mechanism of natural selection to explain it.

The scientific knowledge of the time was insufficient to provide support for a full-fledged theory of natural selection, primarily because of a lack of understanding of heredity. Although the Austrian monk Gregor Mendel had discovered the basic principles of heredity, he labored in obscurity, his insights unknown to other scientists of his time. How organisms passed information from generation to generation was a puzzle. Many theories involved the idea that some activity of the individual animal caused organic change that was subsequently passed to offspring—by mechanisms only guessed at. Darwin himself favored a blending type of inheritance in which particles ("gemmules") from all parts of the parents' bodies would flow to the reproductive organs, where they would be blended and passed on to the offspring.

But natural selection could not be combined with blending inheritance or various models on which acquired characteristics are inherited because such mechanisms would reduce genetic variation each generation. Natural selection requires variation in a population, and the organisms that have characteristics most suitable to a particular environment are the ones that tend to survive and reproduce. Natural selection thus requires that variation be continually renewed each generation; both blending inheritance and natural selection itself would reduce variation. Adaptation would be unlikely to occur. In Darwin's day, many (though not all) scientists concluded that there were critical problems with natural selection as a mechanism of evolution.

It was not until the early twentieth century that it became clear that variability does not reduce each generation. Gregor Mendel's rediscovered (and confirmed) research on pea plants showed that whatever it was that was passed on from generation to generation (later to be called genes, and even later to be recognized as DNA-encoded instructions), it did not blend in the offspring, but remained separate, even if hidden for one or more generations. Heredity is particulate, not blending. Furthermore, genetic information is shuffled each time a sperm fertilizes an egg. Given the particulate nature of inheritance and the existence of phenomena such as dominance and recessiveness, it was clear that natural selection would have sufficient variation on which to operate.

In the late nineteenth and early twentieth centuries, natural selection nonetheless competed with alternate explanations of evolution (Bowler 1988: 7), including a brief revival in popularity of Jean-Baptiste Lamarck's views of the inheritance of acquired characteristics. Lamarckism pointed to observable change: the activities engaged in by an individual during its life could affect its size, shape, and even other characteristics. If these characteristics could be passed on to its offspring, a mechanism would exist to bring about adaptive change. A rabbit living in a cold climate grew a thicker coat; did it pass on its thicker coat to its offspring? There seemed to be evidence of such things: the blacksmith developed large muscles, and the blacksmith's son also tended to be well-muscled—but was this a result of the blacksmith's passing down the big muscles acquired swinging a hammer at the forge? Or was there another explanation, such as the son's going into the family business (and having inherited the potential to develop large muscles under conditions of strenuous exercise)? Without a better knowledge of how heredity operated, evolution by natural selection looked no more plausible than Lamarckism and other teleological explanations.

In the 1890s, the German biologist August Weismann performed an experiment that was instrumental in convincing most scientists that Lamarckian evolution was untenable. First, he cut the tails off of a number of rats, and then bred them to one another. When the rat pups were born, all of them had normal tails; so he cut them off and again bred the offspring with one another. The next generation of rats were also born with normal-length tails. Weismann continued his experiment for 20 generations of rats, and in each and every new generation, there was no inheritance of the acquired trait of cropped tails. The combination of reduced confidence in Lamarckism together with experimental demonstration of Mendelian principles of heredity moved Mendelian genetics to the forefront during the 1930s.

In the 1940s, Darwinian natural selection and Mendelian genetics came together as scientists recognized the powerful support Mendelian genetics provided to the basic Darwinian model of evolution by natural selection. Called the "neo-Darwinian synthesis," or "neo-Darwinism," it remains a basic approach to understanding the mechanisms of evolution. Neo-Darwinism further has been expanded by the second genetic revolution of the twentieth century, the discovery of the molecular basis of heredity. Since the 1953 discovery by James Watson and Francis Crick of the structure of DNA, the hereditary material of cells, investigation of the molecular basis of life has expanded almost exponentially to become perhaps the most active—and certainly the best-funded—area of biological research. Such knowledge has also informed our understanding of the relationships among living things. The "big idea" of descent with modification—that the more recently two forms have shared a common ancestor, the more similar they will be—is reflected not only in anatomy and behavior but also in proteins.

It is safe to say that by the mid-twentieth century, mainstream science in both Europe and the United States was unanimous not only on common ancestry of living things but also on natural selection's being the main—though not the only—force bringing about evolution. The late twentieth century advances in biochemistry and molecular biology have further substantiated these conclusions.

Darwin's Science

In addition to the idea of evolution by natural selection, *On the Origin of Species* illustrated a somewhat different way of looking at biology and a different philosophy of science from that familiar to Darwin's contemporaries (Mayr 1964: xviii).

A *New Conception of Biology*. For Darwin, transmutation of species was a natural phenomenon: it neither required a guiding hand nor resulted in a predetermined goal. Species changed as a result of the need to adapt to immediate environmental circumstances. Since the geology of the planet, and thus environmental circumstances, changed over time, there could not be an ultimate goal toward which creation was heading. It was not possible to predict future changes in living organisms. Darwin's view of science restricted scientific explanations to natural cause. In this he was preceded and influenced by changes that had taken place during the previous 100 years or so in the field of geology (Gillespie 1979: 11).

In the late 1700s, the Scottish geologist James Hutton proposed a view that became known as uniformitarianism: that Earth was ancient, and that its surface could be explained by processes we see taking place today—sedimentation, erosion, faulting, flooding, and the like. There was no need to invoke the direct hand of God to explain the building up of mountains, the presence of seas modern or ancient, or the accumulation of layers of strata. Geology could be understood through natural processes. Darwin's mentor and friend Charles Lyell promoted uniformitarianism in the 1830s and beyond, and the view came to predominate—though not without opposition.

Uniformitarian geologists eventually won the day, but biologists lagged behind; a seminal uniformitarian text by the Scottish scientist John Playfair, *Illustrations of the Huttonian Theory*, was published in 1802—the same year that William Paley published his argument for design, *Natural Theology*. But the seeds for a naturalistic foundation for biology had been planted: geology, after all, has consequences for biology, since the geological column is partly defined by fossils. Different strata are regularly marked by the disappearance of some life-forms and the appearance of new ones (even if they are similar to previous ones). How can these be explained? Creationist geologists required that God re-create life forms after every catastrophic geological change. Darwin viewed the appearance of new species in a stratum as the result of evolutionary change, of descent with modification from earlier ancestors. His mechanism of natural selection likewise reinforced the conclusion that the fossil record and current diversity of life could be explained without recourse to divine intervention. Darwin's bold naturalism applied to biology proved difficult for many critics to take. Many scientists and theologians objected to Darwin's removal of the need for divine intervention in the biological sciences—much as critics of uniformitarian geology had protested a century before.

Similarly, because there was so much evidence that species had indeed changed through time, and because Darwin's and other scientists' studies of both wild and domesticated animals and plants had demonstrated great variation of form within species, equally untenable were typological species concepts in which species were conceived of as reflections of a Platonic *eidos*. Darwin practiced what the modern biologist Ernst Mayr calls "population thinking," where the object of study of biology is actual individual-to-individual variations rather than an abstract concept of an ideal form.

Perhaps because Darwin was fundamentally a naturalist with broad knowledge of living plants and animals, he was able to conceive of species as having almost unlimited variation, which allowed him to speculate about variation as a source of gradual adaptation and eventual transmutation.

A New Conception of Science. The expectation of scientists in the mid-nineteenth century was that the goal of science was the accumulation of certain knowledge. A successful scientific explanation resulted in positive finality. Anything less than certitude was deficient (Moore 1979: 194).

According to this Baconian inductivist approach (derived from the writings of the sixteenth-century scholar Francis Bacon), the scientist who properly performs his craft is one who patiently collects facts, assembles them in a logical and orderly fashion, and lets explanations arise out of this network of ideas. "The outcome of repeated inductions would be a series of propositions, decreasing in number, increasing in generality, and culminating in 'those laws and determinations of absolute actuality' which can be known to be certainly true" (Moore 1979: 194; internal quote from John Losee, *A Historical Introduction to the Philosophy of Science*. London: Oxford University Press,

1972, pp. 164–167). A scientific explanation was considered to have been proven when it accounted for all the facts, and thus was a complete and certain law of nature.

Of course, such an ideal is hardly ever obtainable. It is the nature of science that new discoveries cause us to rethink our conclusions and rework our explanations. Today no one thinks that there is ever final certainty to a scientific explanation, but in the late eighteenth century, such a view was common—though not universal—among scientists and other educated individuals. This was not Darwin's approach, however.

Darwin recognized that the world is not static and, with his abundant knowledge of natural history, knew that variability characterizing natural phenomena would make the certainty sought by the Baconian approach highly improbable. How could one account for all the facts if new facts were continually being generated? "The lesson was plain: induction, no matter how rigorous, could never rule out the possibility of alternative explanations" (Moore 1979: 196).

Darwin's approach to science indeed was to collect facts (and there is an abundance of them in the *Origin*—Darwin was a skilled natural historian and experimenter)—but to collect them with a hypothesis or tentative explanation in mind. Those hypotheses that were not factually disproved were used to generate additional hypotheses, which were themselves tested against the facts. A network of inferences thereby was established. Darwin was careful to state how his hypotheses and generalizations could be tested by listing what sort of observations would have to be made to disprove his views—but he also firmly asserted that until that time, his explanations were the best available.

Rather than the Baconian approach of presenting his views when they were "proved" or "certain," Darwin's approach was to present a coherent set of supported inferences, arguing that the lack of counterevidence gave them the highest probability of being an accurate or true explanation. The probabilistic approach to science, reflecting a dynamic universe, was a sharp contrast to the older Baconian approach of many of Darwin's contemporaries, many of whom viewed the universe as specially designed and largely static. According to Moore, Darwin's approach to science itself was one of the major reasons that the concept of evolution by natural selection presented in *On the Origin of Species* (Moore 1979) was rejected. Darwin's great work was denounced as being "speculative," "probabilistic," "unsupported," and "far from proven." Yet Darwin's way of doing science—probabilities and all—is much more familiar to us in the twenty-first century than that of his contemporaries.

The Religious Response to *On the Origin of Species*

Christians who reject evolution tend to reject it for one or both of two reasons. Common descent conflicts with biblical Special Creation. The Bible in one literal reading tells of the universe's creation in six days, yet data from physics, astronomy, geology, and biology support a picture of the universe unfolding over billions of years. First there is the big bang, then gas clouds, then stars and galaxies, and only about 4 billion years ago does planet Earth form. Life does not appear for another 2 billion years or so, and then not all at once (see chapter 2). The Bible reading literally also suggests that this creation event occurred a relatively short time ago, geologically speaking—a span measured over thousands rather than billions of years. Yet data from physics and geology firmly support the inference that Earth is ancient. A literal reading of Genesis has animal "kinds" appearing in their present form, and varying only "within the kind," whereas biology, genetics, and geology strongly support the inference that species change through time. The perspective of Special Creationism is that of a sudden, recent, unchanging universe, whereas the perspective of evolution is that of a gradually appearing, ancient, changing universe. It is not surprising that two such different perspectives clash.

Modern mainstream Christians generally are not biblical literalists and thus do not regard the incompatibility of evolution with biblical literalism as a reason to reject the former. Not believing in created kinds, they have no theological objection to living things descending with modification from common ancestors. But there is a second reason that Christians reject evolution, shared by literalists and nonliteralists alike, and this is the issue of design, purpose, and meaning.

The Problem of Design and Purpose. In Aristotelian philosophy, the purpose or end result of something is thought to be a *cause*. Explaining something by its purpose is known as *teleology*. Up until the nineteenth century, the cause of the marvelous wonders of nature, including the intricacies of anatomical structure, was widely considered to be God's purposive design. Thus the fit of an organism to its environment was the result of the special creation of its features.

While a college student, Darwin read works by the theologian William Paley, whose *Natural Theology; Or Evidences of the Existence and Attributes of the Deity, Collected from the Appearances of Nature* he thought a splendid book (Paley 1803). Paley's view was that God specifically designed complex structures to meet the needs of organisms. *Natural Theology* was also an apologetic, or religious proof of the existence of God; Paley's version of the Argument from Design is considered a classic. God's existence could be proved, said Paley, by the existence of structural complexity in nature. In a famous analogy, he compared finding a stone on a heath to finding a watch. The former could have been there forever; it was a natural object and did not require any special explanation. But the watch was obviously an artifact—its springs, wires, and other components were assembled to mark the passage of time. Structural complexity that achieved a purpose was evidence for design and therefore of a designer. When we see a natural structure such as the vertebrate eye, which accomplishes the purpose of allowing sight, we can similarly infer design and hence a designer. The existence of structures such as the vertebrate eye is evidence for the existence of God, according to this analogy.

Paley contrasted design with chance, and it was clearly as absurd to believe that something like the vertebrate eye could assemble by chance as it was to believe that the parts of a watch might come together and function due to random movements of springs and wires. Modern creationists take the same view, equating evolution with chance (in the sense of being unguided and purposeless) and contrasting it with guided design. A favorite creationist argument is quite Paleyean: many cite astronomer Fred Hoyle's estimate of the possibility of life forming "by chance" as equivalent to a Boeing

707 airplane's being assembled by a whirlwind passing through a junkyard.

But even before Darwin's *Origin*, the Argument from Design was proving to be less useful in understanding the natural world. Part of this was due to increased knowledge of the natural world during the 1700s and early 1800s. As naturalists examined the world and its creatures more carefully, it became clear that William Paley's ideas of the perfection of structural complexity didn't match reality. Although there were many wonderful structures that admirably suited organisms to their environments—waterproof feathers of ducks, or the hollow bones of birds that provide strength with lightness—there were also curious constructions that didn't seem to make survival more probable, like reduced wings in kiwis and similar flightless birds. Other structural oddities seemed unnecessarily complex, such as the migration of the eyes of young flounders from a normal position on either side of the head to both eyes on one side of the head. If flounders are to be adapted to living flat on the ocean floor, why are they not born with both eyes on the same side of the head? Examples can be multiplied (for examples from a modern author, see Gould 1980), but the point was recognized even before Darwin that there were many examples of odd structures that didn't appear to have been the direct creation of an omniscient, benevolent God. The weight of natural historical observations was weakening the Argument from Design.

Natural selection, of course, provided a natural means to explain the origin of complex structures that adapted their owners to their environments. As discussed in chapter 2, those organisms having structures that better suited them to a particular environment were more likely to leave descendants than those lacking these useful structures. Populations would thus change over time, their members becoming better adapted to their environments. Paley was correct to choose design over chance, but he did not know that there was a natural as well as a transcendent source of design.

But how to choose between transcendent design and natural selection? Obviously an omniscient Creator could specially create structures such as the vertebrate eye—but so could natural selection. Either the direct hand of God or natural selection could explain well-designed structures. In fact, in the *Origin*, Darwin used Paley's example of the vertebrate eye to illustrate how a complex structure might plausibly result through natural selection. More difficult for the Argument from Design was explaining those structures that just barely worked, or that were obviously cobbled together from disparate parts having other functions in related species. Natural selection can operate only on available variations, so if the "right" variation is not available, either the population dies out or some other structure will have to be modified into an adaptation. So nature is full of oddities like antennae modified into fishing lures, or jaw bones turned into hearing structures—things that don't look so much engineered as tinkered with (Jacob 1977).

Along the same lines, some structures seem to barely work, but because natural selection is the "survival of the fit enough"; it is not expected that "perfect," optimal structures will always be the end result. Thus natural selection can account for both well-designed (in the sense of good or efficient operation) as well as poorly designed structures. On the other hand, for God to have deliberately created jerry-rigged, odd, or poorly designed structures is of course possible, but it is theologically unsatisfying and empirically untestable. Natural selection, in fact, offered a theological way out to those concerned with this issue: God could work through natural selection and thus not be stuck with accusations of bad design.

The power of natural selection to explain the oddities of nature drew people away from design as a scientific explanation. It became possible to explain structural complexity and adaptation through natural cause. Still, there remained a theological problem: if Darwin was right, and natural selection explained design, the implication was clear: God did not need to create humans directly. But if God did not create humans directly, did this mean that humans were less special to God? Traditional teleological views held that humans existed because God created them with a specific purpose. If humans were the result of a natural process that didn't *require* the direct involvement of God, did that negate an ultimate meaning or purpose to life? (Chapter 11 presents theological responses to this question.)

Both biblical literalism and problems with design and purpose played roles in the reception of Darwin's ideas in the nineteenth century. In the early nineteenth century, both arguments were raised against evolution, at a time when links between science and religion were still strong.

Science and Religion. In the United States, early nineteenth-century religious intellectuals (clergy, theologians, and religious scientists and laymen) embraced science as providing proof of design, the existence of God, and other Christian theological positions. Many nineteenth-century scientists worked within a theological framework and frequently referred to religious views in discussing scientific positions. "The existence of God, the reality of His providential concern for his creation, the veracity of miracles, the importance of humanity as the focus of divine plan—all these doctrines appeared to be legitimate inferences from the clearest disclosures of scientific investigation" (Roberts 1988: 13).

As geologists explored the fossil record in the early half of the century, the sequence of changing forms through time was seen to reflect separate creations—and progressively improved ones, as well. This, too, harmonized with the Christian view that there was a divine providential plan unfolding through time. That these now-extinct creatures were also adapted to their environments reinforced the Argument from Design. It was, however, a time of rapid growth of scientific knowledge, and these same geological observations encouraged an alternative explanation: the transmutation of species.

Science itself was evolving into a more naturalistic methodology, as natural explanations provided more testable and reliable inferences than supernatural causes. One of the early presentations of the idea of transmutation of species appeared in Robert Chambers's *Vestiges of the Natural History of Creation*, published anonymously in 1844—just about the time Darwin was beginning to work in earnest on the principle of natural selection. In Chambers's view, living things adapted to their environments in response to God-created law, rather than being specially created for that purpose. His hands-off, non-miracle-generating Creator was widely rejected by many clerics because it did not reflect the personal God with which they were familiar (Roberts 1988). Furthermore, scientists were unimpressed by the somewhat wispy scientific mechanisms Chambers proposed.

By midcentury, then, transmutation and changes in ideas of how science should be done were in the air. Darwin's science pushed the boundaries much farther than did Chambers, and *On the Origin of Species* subsequently experienced an even stronger reaction from the religious community. But the seeds of change had been sown: the concept of a dynamic rather than a static world, already accepted in astronomy and growing in geology, would eventually wash over biology as well. But if the universe were in a state of change, of evolution, did this negate the Christian view that there was an overarching purpose for the universe? For humankind? There were many theological issues affected by Darwin's views, some of which are still being grappled with today.

Religious Responses in Context. It is not easy to summarize the religious response to Darwin's ideas. According to Roberts, there was no pressure on clerics to modify traditional views until the scientific community coalesced around transmutation; previous theories of species change such as Chambers's had been rejected by the scientific community, obviating the need for the theological community to grapple with contradictions between the new science and traditional Christianity. However, by the mid-1870s, the scientific community became convinced of transmutation, and religious leaders were forced seriously to consider evolution (Roberts 1988). The initial reaction of British clergy to Darwin's ideas was mixed, but eventually the strength of the science and the need for coherence between science and theology brought about sufficient accommodation.

Still, the process took decades, and in some denominations, resolution was not achieved until the twentieth century. The acceptance of Darwinian ideas was neither uniform nor simple, reflecting not only the growing tension between religion and science but also local issues that at some times and places were more important than the scientific or theological ones. Even within a religious tradition, there could be significant differences in the degree to which evolution was accepted. In Presbyterian Edinburgh, for example, during the 1870s evolution was generally accepted by the leading clerics, who were more concerned with the growth of German Modernism and biblical criticism and its consequences for Presbyterian theology. In the late 1870s and 1880s, Presbyterians in the United States, led by Princeton University theologian Charles Hodge, generally accepted evolution but rejected Darwin's mechanism of natural selection. On the other hand, in Belfast, Ireland, Presbyterians spoke out strongly against both evolution and the natural selection mechanism (Livingstone 1999).

The Catholic response to evolution was similarly complex, reflecting social movements and consequent church politics over the struggle of American Catholics to define a Catholic identity that reflected their national and cultural needs. The late nineteenth and early twentieth centuries were periods of increased immigration to America by European Catholics, primarily from Ireland, Germany, Italy, and Poland. Progressive American Catholic leaders were eager to integrate the largely working-class newcomers into American society, to educate and Americanize them, which would reduce their marginality. The presence of so many foreign Catholics, "different" from the Anglo-Protestant majority, tended to marginalize the Catholic religion as well. Americanizing the newcomers would also Americanize Catholicism, and vice versa. Progressive Catholics sought to define Catholicism in terms of American tradition and history. However, "Americanism," which included the separation of church and state, support of labor unions (important to immigrants), individual liberty, and material progress as well as spiritual, was seen as a threat by the Vatican (Appleby 1999). Progressive Catholics promoting Americanism were also more accepting of biblical criticism, science, and evolution, thus tainting evolution in the eyes of the Vatican. The fact that liberal Protestants largely accepted evolution made evolution even less palatable to conservative Catholics.

Turn-of-the-century Catholic immigration generated a nativist backlash from the Anglo-Protestant majority, and this, too, became entangled with attitudes toward evolution. Evolution and natural selection were incorporated into the anti-immigrationist arguments and also into the subsequent eugenics movement of the early twentieth century. Thus evolution became associated in some minds with anti-Catholicism, sterilization of the "feeble-minded" (read: Catholic immigrants), birth control, and racism (Appleby 1999). To be sure, there was doctrinal opposition to evolution, but because the acceptance or rejection of evolution was so embedded in other issues, it made it easier for the Catholic Church to eventually accept evolution.

> First, the debate over Darwinism itself, for all its virulence, was actually an occasion for American Catholics to work out a number of identity-defining issues facing the immigrant community. Second, the debate did not leave a strong antievolutionist legacy to future Catholic educators in quite the way Protestant fundamentalism did. The advent of evolutionary theory, in other words, served as a catalyst for the resolution of internal Catholic issues rather than as a sustained evaluation of Darwinism and evolutionary theory in itself. Despite the seeming victory of the conservatives, moreover, the scientific theory of evolution was never formally condemned, and Roman Catholicism modified its general anti-evolutionist stance several times in the twentieth century; this was a process that culminated in the conditional approval of the theory by Pope Pius XII in 1950. (Appleby 1999: 179)

In addition to Pius XII, subsequent popes have offered an accommodation of Catholic theology to science along the lines sketched in the nineteenth century by the Catholic scientist St. George Mivart: the human soul is directly infused by God but the body has evolved from animal predecessors. Catholic high schools thus routinely teach evolution, as it has no formal doctrinal conflict with Catholic theology.

In both the United States and Great Britain, religious objection to evolution was spurred on by the anticlericalism of some of Darwin's early defenders, especially Thomas Henry Huxley and Herbert Spencer. It was easy for religious intellectuals to reject evolution by natural selection when it was presented by some of its supporters as compelling atheist belief. The active support of evolution in the 1860s and 1870s of a number of American scientists who were also active churchmen, such as Asa Gray, greatly helped to defuse the idea that evolution was an inherently atheistic idea (Numbers 1998).

By mid-twentieth century in Great Britain, Europe, and North America, the scientific community no longer questioned whether evolution occurred. The neo-Darwinian revolution of the 1930s and 1940s had been successful (see chapter 3). In Great Britain and Europe, but not in North America, evolution was included matter-of-factly in textbooks and curricula of education systems. In the United States, however, evolution was a topic consistently taught only at the college level—being largely absent from the K–12 curriculum. Understanding this difference requires looking more closely at American history.

BACKGROUND TO CONFLICT

Why was evolution absent from American schools in the early twentieth century? To understand, we need to reflect upon both American religious history and the educational structure of the United States.

America's Decentralized Educational System

Consider the settlement history of the United States: beginning with Northern European (English, Dutch) contact in the northeastern part of the continent and Southern European (Spanish, Portuguese) exploration of the south and west, the movement of people began at the continental coasts and worked inward. After the initial trappers and explorers mapped out the territory, settlers filled in the river valleys, using the vast interior waterways as arteries for trade and communication. People preceded government: territorial or state governmental services we today take for granted, such as police, health care, and maintenance of public facilities such as roads and bridges, usually lagged well behind the expansion of people into new territories. The contributions of state or territorial governing bodies were rarely felt; hardly ever were federal agencies functional in the early settlements. This in fact paralleled the experiences of the earliest European settlers, deposited with no support from their governments on the shores of a new land—which they more often than not must have viewed with very mixed feelings of both opportunity and foreboding as the ships that had brought them sailed back to civilization.

Because of this lack of connection with government agencies, and the independent structure of states relative to the national government, frontier communities were generally responsible for setting up their own school systems largely independent of state and federal agencies. Local communities determined whether there would be a school, constructed the building—if there was one—and determined who should teach, what he or she would be paid, and even the content of what the teacher would teach. Local control began as a necessity, and through custom became enshrined as a right.

To this day, American education remains remarkably decentralized. The federal government has a role to play in education, but it is dwarfed by the responsibility and activity of states and local school districts. In some states, a large percentage—even a preponderance—of the budget is devoted to education, and states rigidly insist on their right to determine the structure and content of the educational system, with a minimum of interference from the federal government. There is a similar tension between most state governments and local school districts. These local districts—which may be cities, regions incorporating more than one city, or smaller units corresponding to neighborhoods or other subdivisions of cities—are governed by locally elected school boards consisting of interested citizens who may or may not know much about the field of education but who, by virtue of being from the community, maintain a localized focus on education. Many states have state-level education standards that are used to guide curricular development in local school districts, but in most states, the districts have the final say as to how much of the state standards in a given field will be utilized in their schools.

For decades, as more money for education has come out of Washington, the U.S. government has argued that it has the right to oversee how its money is spent, though the federal government has been quicker to stress financial accountability than academic content. In 1989, the first Bush administration's Department of Education proposed the establishment of national standards for history, mathematics, and science—but, reflecting the emphasis in American education on local control, such standards were to be only advisory, not mandatory. The National Science Education Standards were published seven years later (National Committee on Science Education Standards and Assessment, National Research Council 1996).

The decentralization of American education is a source of wonder to Europeans and the Japanese, for example, who have curricula that are uniform across all communities in their nations. In France, for example, the curriculum in any particular grade is virtually the same from week to week in any classroom in any city. In the United States, even schools within the same district may not teach the same subjects in the same order, or even in the same year.

America's Decentralized Religious History

American religious history reflects an equally decentralized, "frontier" orientation. The nation initially was settled largely by religious dissidents, who came here at least partly for their own religious freedom—though they generally discriminated against people practicing other faiths! The first East Coast settlers were mostly Protestant and generally came from congregational traditions in which most decisions were made at the level of the individual church, rather than imposed hierarchically from church bureaucracies. The nature of the frontier reinforced this tendency: pioneers establishing new settlements had to establish not only police and educational systems but also churches, if they wanted them: certainly the government was not going to do so. As a result, churches took on a regional flavor, often diverging theologically from other churches nominally the same.

The United States also has been the nursery for a wide variety of spontaneously generated, independent sects, often inspired by charismatic leaders. It was in the United States that the Seventh-Day Adventists, the Church of Jesus Christ of Latter-Day Saints (Mormons), Jehovah's Witnesses, Christian Scientists, and now-extinct sects such as Shakers and Millerites were founded, reflecting our decentralized, nonhierarchical religious past. But perhaps the most important reason that modern antievolutionism developed here rather than in, say, Europe, was the founding in 1910–1915 of Fundamentalism, a Protestant view that stresses the inerrancy of the Bible. Fundamentalism was not successfully exported to Europe or Great Britain, but it formed the basis in the United States for the antievolutionism of the 1920s Scopes era as well as the present day.

AMERICAN ANTIEVOLUTIONISM

Antievolutionism in the United States can be divided into three periods. During the first, antievolutionists worked to pass legislation that would eliminate evolution from the classroom and textbooks. When laws restricting the teaching of evolution were eventually struck down, Creation Science developed, bringing about the second major period of American antievolutionism. These two periods are the subject of the next chapter. When laws promoting equal time for Creation Science were eventually struck down, antievolution forces regrouped under a diverse set of schemes, including various repackagings of Creation Science as well as some new offerings. Chapter 6 will describe these changes.

REFERENCES CITED

Appleby, R. Scott. 1999. Exposing Darwin's "Hidden Agenda": Roman Catholic Responses to Evolution, 1875–1915. In *Disseminating Darwinism: The Role of Place, Race, Religion, and Gender*, edited by R. L. Numbers and J. Stenhouse. Cambridge: Cambridge University Press.

Bowler, Peter J. 1988. *The Non-Darwinian Revolution: Reinterpreting a Historical Myth*. Baltimore: Johns Hopkins University Press.

Burkhardt, Frederick, and Sydney Smith, eds. 2002. *The Correspondence of Charles Darwin*. Vol. 13, *1865*. Cambridge: Cambridge University Press.

Durant, John R. 1998. A Critical-Historical Perspective on the Argument About Evolution and Creation. In *An Evolving Dialogue: Scientific, Historical, Philosophical and Theological Perspectives on Evolution*, edited by J. B. Miller. Washington, DC: American Association for the Advancement of Science.

Gillespie, Neil C. 1979. *Charles Darwin and the Problem of Creation*. Chicago: University of Chicago Press.

Gossett, Thomas F. 1965. *Race: The History of an Idea in America*. New York: Schocken Books.

Gould, Stephen Jay. 1980. The Panda's Thumb. In Gould's *The Panda's Thumb: More Reflections on Natural History*. New York: Norton.

Jacob, Francois. 1977. Evolution and Tinkering. *Science* 196 (4295): 1161–1166.

Livingstone, David N. 1999. Science, Region, and Religion: The Reception of Darwinism in Princeton, Belfast, and Edinburgh. In *Disseminating Darwinism: The Role of Place, Race, Religion, and Gender*, edited by R. L. Numbers and J. S. Stenhouse. Cambridge: Cambridge University Press.

Mayr, Ernst. 1964. Introduction. In *On the Origin of Species by Charles Darwin: A Facsimile of the First Edition*. Cambridge, MA: Harvard University Press.

Moore, James R. 1979. *The Post-Darwinian Controversies: A Study of the Protestant Struggle to Come to Terms with Darwin in Great Britain and America 1870–1900*. Cambridge: Cambridge University Press.

National Committee on Science Education Standards and Assessment, National Research Council. 1996. *National Science Education Standards*. Washington, DC: National Academy Press.

Numbers, Ronald. 1992. *The Creationists*. New York: Knopf.

Numbers, Ronald L. 1998. *Darwinism Comes to America*. Cambridge, MA: Harvard University Press.

Paley, William. 1803. *Natural Theology: Or, Evidences of the Existence and Attributes of the Deity, Collected from the Appearances of Nature*, 5th ed. London: Faulder.

Roberts, Jon H. 1988. *Darwinism and the Divine in America: Protestant Intellectuals and Organic Evolution, 1859–1900*. Madison: University of Wisconsin Press.

Winchester, Simon. 2001. *The Map That Changed the World: William Smith and the Birth of Modern Geology*. New York: HarperCollins.

CHAPTER 5

Eliminating Evolution, Inventing Creation Science

A GROWING CRISIS

As discussed in chapter 4, evolution became well-accepted by the scientific community by the turn of the twentieth century. It thereafter began to be included in college and secondary school textbooks. Partly because of the efforts of American scientists who accepted evolution and who also were active church members, the late nineteenth century was not a period of extensive religious hostility to evolution. It was not until the twentieth century that the antievolution movement became organized, active, and effective. Three trends converged to produce the first major manifestation of antievolutionism in the twentieth century: the growth of secondary education, the appearance of Protestant Fundamentalism, and the association of evolution with social and political ideas of social Darwinism that became unpopular after World War I.

Although textbooks at the turn of the century included evolution, few students were exposed to the evolution contained in these books: in the late nineteenth century, high school education was largely limited to urban dwellers and the elite. In 1890, for example, only 3.8 percent of children ages 14–17 attended school—about 202,960 students (Larson 2003: 26). But high school enrollment approximately doubled during each subsequent decade, so that by 1920, there were almost 2 million students attending high school. The practical effect of this was that more students were being exposed to evolution—and parents who felt uneasy about evolution for religious or political reasons rallied around the politician William Jennings Bryan to protest the teaching of evolution to their children.

Fundamentalism

The Fundamentalist movement in American Protestantism is named for a theological perspective developed during the first few decades of the twentieth century. It was encapsulated in a series of small booklets collectively called *The Fundamentals*, published between 1910 and 1915 (Armstrong 2000: 171). Its roots, however, go back to earlier conservative Protestant movements. Fundamentalism is partly a reaction to a theological movement called Modernism that began in Germany in the 1880s. Modernism reflected a technique of biblical interpretation called Higher Criticism, which proposed looking at the Bible in its cultural, historical, and even literary contexts. Creation and Flood stories, for example, were shown by comparison of ancient texts to have been influenced by similar stories from earlier non-Hebrew religions. With such interpretations, the Bible could be viewed as a product of human agency—with all that suggests of the possibilities of error, misunderstanding, and contradiction—as well as a product of divine inspiration.

More conservative Christians preferred a more traditional interpretation on which the Bible was considered *inerrant* (wholly true and free from error—though some individuals qualified inerrancy as applying only to the version of the Bible God gave to the original authors). Passages were to be taken at face value when at all possible, rather than being "interpreted." Fundamentalists stressed "(1) the inerrancy of Scripture, (2) the Virgin Birth of Christ, (3) Christ's atonement for our sins on the cross, (4) his bodily resurrection and (5) the objective reality of his miracles" (Armstrong 2000: 171).

Financed by millionaires who had founded a conservative Baptist university in Los Angeles (Bible Institute of Los Angeles, now Biola), millions of copies of *The Fundamentals* booklets were printed and distributed "free of charge, to every pastor, professor, and theology student in America" (Armstrong 2000: 171). Different essays treated evolution in different ways: some of the authors rejected evolution, but some accepted various forms of Theistic Evolution. In some, natural selection was rejected, but not common ancestry itself (Larson 1997). Some writers allowed for animal evolution, but not human, and some even allowed for human evolution, though not through natural selection. Natural selection was opposed because it replaced God's direct action with natural causes and thus indicated to some a less personal, hands-on, involved God, which Fundamentalist theology rejected. Most of the authors of *The Fundamentals* were Day-Age Creationists, allowing for an old Earth but insisting on a recent appearance of humans. Although not all the *Fundamentals* booklets were antievolutionary, the Fundamentalist position hardened toward evolution fairly quickly. Fundamentalists became the ground troops for the campaign to rid schools of evolution. They were motivated by religious sentiments and also by a concern that evolution was the source of many negative and even corrosive social trends.

Evolution as Social Evil

The second decade of the twentieth century was a time of considerable social and psychological unrest. The appalling death, brutality, destruction, and devastation of World War I led many citizens, including many conservative Christians, to conclude that civilization itself had failed. Conservative Christians sought a solution in a return to biblical authority and in the literal interpretation of Scripture. Their views were further reinforced by Germany's having been the main source of both Higher Criticism, viewed as an attack on religion, and World War I militarism, viewed as an attack on civilization (Marsden 1980; Armstrong 2000).

German militarism, theories of racial superiority, and eugenics were felt by conservative American Christians to be directly related to the acceptance of evolution by Germany at the end of the nineteenth century. In reality, German views of evolution were quite different from those of Darwin, largely rejecting natural selection as a mechanism of change, biological or societal. Evolution by natural selection did not fit German militaristic views of the inevitability of Teutonic triumph; natural selection relies upon selection of the "most fit" in terms of a particular environment. It does not support the idea that Germans or anyone else inevitably would be superior to all others, regardless of environmental circumstance.

In the early twentieth century, evolution was also "credited" with providing the foundation for laissez-faire capitalism, as robber barons of the late nineteenth and early twentieth centuries sometimes claimed that natural selection justified their exploitative labor policies and cutthroat business practices:

> The price which society pays for the law of competition, like the price it pays for cheap comforts and luxuries[,] is also great; but the advantages of this law are also greater still for it is to this law that we owe our wonderful material development, which brings improved conditions in its train. But whether the law be benign or not, we cannot evade it; no substitutes for it have been found, and while the law may be sometimes hard for the individual, it is best for the race, because it insures the survival of the fittest in every department. (Carnegie 1889: 653)

Thus Fundamentalists, led by the famous progressive politician and champion of the workingman William Jennings Bryan, had many reasons to oppose the teaching of evolution to their children, whether or not these reasons were justified. Beginning in the early 1920s, several state legislatures took up Bryan's call to outlaw evolution, and finally, on March 23, 1925, Tennessee passed the Butler Act. This set in motion events that would culminate in the "Trial of the Century."

SHOWDOWN IN DAYTON

"It shall be unlawful for any teacher to teach any theory that denies the Story of Divine Creation of man as taught in the Bible, and to teach instead that man has descended from a lower order of animal" declared Tennessee's Butler Act (Larson 2003: 54). It was passed by the House with virtually no debate, but the Senate heard considerable testimony both for and against it. Scientists almost uniformly opposed it, and the religious community was split, with Fundamentalists strongly supporting the bill as a means of preserving children's faith, and liberals opposing it on the grounds that the state should not favor one religious position over another. Public sentiment in Tennessee was so strong, though, that the state's senators felt great pressure to pass the bill, and did.

Almost immediately, the young American Civil Liberties Union in New York took up the challenge of testing the new law. Because of restrictions on civil liberties imposed by the government during and after World War I, the ACLU was particularly concerned with free speech. The early 1920s and the preceding decade were a time of social unrest, economic insecurity, and agitation for worker rights. Strikes in mills and mines resulted in repression of labor. Many Americans feared that European anarchism and socialism were taking root in American soil. The ACLU focused on the free speech and other rights of workers—and schoolteachers qualified as labor. ACLU leaders believed that the Butler Act infringed on the free speech rights of teachers by restricting what they could teach.

The ACLU took out advertisements in Tennessee newspapers, offering to defend any teacher willing to volunteer to be the plaintiff in a legal challenge to the Butler Act. Businessmen in the small town of Dayton concocted a plan to bring publicity and business to their community as the site of a high-visibility trial challenging a controversial law. They persuaded John T. Scopes, a young science teacher, to be the ACLU's test case. He would be accused of teaching evolution, the trial would be held, the law would be struck down, and Dayton would receive publicity and a welcome economic shot in the arm. The scenario played out almost as planned.

Scopes was a young man of 24 who taught science at the high school. As the tale is told, town leaders called him in from a tennis game to pitch the idea of challenging the Butler Act in Dayton.

> Scopes presented the ideal defendant for the test case. Single, easy-going, and without any fixed intention of staying in Dayton, he had little to lose from a summertime caper—unlike the regular biology teacher, who had a family and administrative responsibilities. Scopes also looked the part of an earnest young teacher, complete with horn-rimmed glasses and a boyish face that made him appear academic but not threatening. Naturally shy, cooperative, and well-liked, he would not alienate parents or taxpayers with soapbox speeches on evolution or give the appearance of a radical or ungrateful public employee. Yet his friends knew that Scopes disapproved of the new law and accepted an evolutionary view of human origins. (Larson 1997: 91)

The amiable Scopes agreed, a warrant was sworn out, and Scopes was duly charged with the crime of violating the Butler Act, after which he returned to his tennis game. Plans were made to hold the trial in Dayton, the seat of Rhea County.

The plan to bring publicity to Dayton succeeded beyond the businessmen's wildest expectations, and certainly beyond what the young schoolteacher had anticipated. The 1925 trial was truly the "Trial of the Century," being the first trial to be covered not only by the print media but also through live radio broadcasts. The trial would have received a lot of attention on its own merits: the Butler Act had received national publicity, and already battle lines had been drawn over the merit of passing antievolution laws. The unexpected appearance of two political giants of the day, William Jennings Bryan for the prosecution and Clarence Darrow for the defense, only heightened public interest. All these factors transformed the trial into a three-ring circus.

Bryan was one of the nation's most famous and popular public figures. He had been the Democratic Party's candidate for president three times and had served as secretary of state under Woodrow Wilson. He had made his political reputation as a well-known promoter of Progressive causes such as women's suffrage, pacifism, and better working conditions for workers. Always a devout man, in his later years he became known as much for his Fundamentalist Christian views as for his Progressivism. Today—largely due to the Scopes trial—much of his political Progressivism has been forgotten. But in the late 1910s and the 1920s, laissez-faire capitalism—the source of poor working conditions, child labor, and worker exploitation—was regarded as tied to beliefs about evolution, which in addition was believed to be antireligious. Bryan's combination of antievolutionism, political Progressivism, and Fundamentalist Christian antievolutionism was not irrational, given the social and political views of the times.

Clarence Darrow was the most famous defense attorney in the country. Like Bryan, he was a political Progressive; he had supported Bryan in the latter's early attempts to become president. Darrow was also a pacifist and a supporter of free speech—which were not uniformly popular positions in the second decade of the twentieth century. He also was a well-known atheist, thus contrasting sharply with Bryan. With two such giants squaring off against one another, the attraction of the trial to the public was irresistible.

The Scopes trial originally had been conceived as a test of the truth of evolution. Both sides—especially the antievolution side—were eager to testify regarding evolution's validity and whether evolution was inherently anti-Christian or led to immoral or unethical behavior. But the prosecution began to lose enthusiasm for this approach when it became clear that the defense was quickly lining up scientists and theologians who would affirm that evolution was scientific and assert that it was not necessarily anti-Christian. The prosecution had a difficult time finding scientists who rejected evolution (Larson 1997: 130). The prosecution then switched its strategy to argue the case narrowly: Did or did not Scopes break the law? Fortunately for the prosecution, once the trial began, the judge quickly took its point of view, ruling that, indeed, the trial would focus only on whether or not Scopes had broken the law (i.e., had taught evolution). The defense's carefully chosen scientific witnesses (from biology, anthropology, and geology) and its three theologians were not permitted to testify.

One of the most memorable moments in the trial involved a daring legal move: Darrow requested Bryan to take the witness stand as an expert on religion. Bryan accepted, against the advice of his co-counsels. He planned to use the opportunity to witness to both supporters and those not yet converted, and to defend Christianity against the atheist Darrow. Unfortunately for him, however, he was made to appear somewhat foolish, as it became clear that he was an expert neither on the Bible nor on comparative religion, and certainly no expert on science or evolution.

Throughout Bryan's examination, Darrow sought to show that certain passages of the Bible cannot rationally be accepted as literally true. Bryan fell for this scheme by admitting that despite his reputation as a promoter of Fundamentalism, he had no explanation for how Joshua lengthened the day by making the sun stand still. Similarly, he could not answer Darrow's questions whether the Noachian flood that allegedly destroyed all life outside the Ark also killed fish, where Cain got his wife, and how the snake that tempted Eve moved before God made it crawl on its belly as punishment. Bryan acknowledged his acceptance of a long Earth history and a Day-Age interpretation of the Genesis account, which of course allowed enough time for evolution to take place. Further undermining his stance against evolution, Bryan confessed that he knew little about comparative religion or science (Anonymous 1990).

Given the narrow grounds on which testimony was allowed, it was a foregone conclusion that Scopes would lose. Both sides anticipated the verdict; to a large degree, the Dayton trial was viewed by both sides as a preliminary step toward the appeals process: eventually the legality of antievolution laws would be tested by the Supreme Court. Scopes was convicted of having taught evolution (though in reality he may never have actually taught that chapter of the textbook). The defense also lost its appeal to the Tennessee Supreme Court: the ACLU's concern that individual freedom should take priority over the government's authority over public employees was rejected in favor of the state's right to set conditions for employment.

In a surprise move, however, the Tennessee Supreme Court then reversed the Scopes conviction on a technicality. The judge (as was not uncommon in such minor cases) had assigned the $100 fine, but the law required the jury to set the penalty. Scopes's conviction was thrown out, which made further appeal moot. The ACLU's plan to appeal the case to the U.S. Supreme Court was thwarted.

AFTEREFFECTS OF THE SCOPES TRIAL

Although antievolution forces prevailed during the Scopes trial, the aftereffects of the "Trial of the Century" were not as clear-cut. Antievolution laws continued to be submitted, but few passed. Mississippi successfully passed an antievolution bill in 1926, but in 1927 bills were defeated in Arkansas, Oklahoma, Missouri, West Virginia, Delaware, Georgia, Alabama, North Carolina, Florida, Minnesota, and California (Holmes 1927). The Supreme Court eventually would have the opportunity to rule on the constitutionality of antievolution laws—but not until the 1960s.

After the Scopes trial, antievolutionism became associated in the popular imagination with conservative religious views—and the most negative stereotypes of such views. Antievolutionists and Fundamentalists in general were portrayed as foolish, unthinking, religious zealots. Particularly effective in contributing to this stereotype were the Dayton dispatches of the acerbic reporter for the *Baltimore Sun*, H. L. Mencken, but accounts written in the 1930s and afterward also reinforced the view that antievolutionism was a campaign of backward (or at best premodern), uneducated, religious fanatics. Although many leaders of the pre-Scopes antievolution movement were from Northern states, after the Scopes trial, antievolutionism became more regionalized, retaining momentum in the South and rural areas of the country, where Fundamentalism remained strong. Where Fundamentalists held political power, regulations were imposed by school boards to restrict the teaching of evolution. But the demographics of Fundamentalism were changing, as it moved from the cities of its origin to the rural South—where it largely disappeared from the view of the mainstream (Eastern, urban) press (Marsden 1980: 184).

Jerome Lawrence and Robert E. Lee were inspired by the issues raised in the Scopes trial in writing their 1955 Broadway play, *Inherit the Wind*. This play and the movies based on it strongly shaped public images of the Scopes trial and contributed to the negative public image of Fundamentalists. Although the authors explicitly distanced themselves from the Scopes trial in the introduction of the play ("It is not 1925. The stage directions set the time as 'Not too long ago.' It might have been yesterday. It could be tomorrow.") and argued that their motivation for writing the play was to consider issues of free speech, the closeness of the story line in *Inherit the Wind* to the events of the Scopes trial was obvious. The play featured a young teacher tried and imprisoned for teaching evolution and thereby violating an antievolution law. Two prominent political figures—one a Fundamentalist and one a freethinker—lined up on the prosecution and defense sides, respectively. Issues of Fundamentalism and Modernism (science) were constantly present. The play even included an H. L. Mencken–like cynical reporter with an acid tongue. The circus atmosphere of the trial in the play certainly paralleled that of the actual trial. The play's strong characters and memorable writing have made it a classic, often read and performed in high schools as a vehicle for discussing issues of free speech as well as the role in society of minority and majority views.

Of course, there were major differences between the Scopes trial and *Inherit the Wind*; the goal of the playwrights was to present a dramatic narrative rather than a historical account. Modern antievolutionists particularly object to the treatment of the character based on Bryan, who is bombastic and a caricature of a religious bigot. Viewed as history, *Inherit the Wind* is clearly inaccurate: although the Scopes-like character in the play goes to jail, Scopes himself was never imprisoned, and a Fundamentalist minister who rails against evolution and science—and who is the father of the young teacher's girlfriend—did not have a Scopes trial counterpart but was added for plot reasons. *Inherit the Wind* was intended, according to its authors, as a metaphor of 1950s McCarthyist politics versus free speech and freedom of conscience.

Perhaps the biggest difference between the picture painted by the play and movies and the actual trial was the false image presented in the fictionalized account that "The light of reason had banished religious obscurantism" (Larson 1997: 246). Neither Fundamentalism nor the antievolutionist campaign disappeared after 1925, though the latter abated somewhat. This was primarily because antievolutionism became largely unnecessary: evolution remained effectively absent from science instruction until the 1960s.

Scopes lost; the antievolution laws remained on the books, and even increased in number. In the South, states and local school districts restricted the teaching of evolution, and teachers and local parents who chose textbooks preferred ones that slighted evolution. The economic pressures were effective: textbook publishers knew they had to remove, downplay, or qualify evolution if they wanted sales, and they did. Books tailored for the Southern markets were of course sold elsewhere, and evolution disappeared from textbooks all over the nation (Grabiner and Miller 1974). Because of the influence textbooks have on the curriculum, with evolution absent from the textbooks, it quickly disappeared from the classroom. By 1930, only five years after the Scopes trial, an estimated 70 percent of American classrooms omitted evolution (Larson 2003: 85), and the amount diminished even further thereafter. Its return sparked the next chapter in American antievolutionism, as creationists lashed back at the reintroduction of evolution in American schools.

CREATION SCIENCE EVOLVES

Even though most consider the Scopes trial a "victory" for evolution, fewer high school teachers taught evolution after the Scopes trial than before. The amount of evolution in textbooks decreased rapidly after 1925 (Skoog 1979; Grabiner and Miller 1974). This remained the case until the late 1950s, when a federally funded campaign to improve precollege science education brought evolution back into textbooks. As evolution eased back into the curriculum, antievolutionists reacted, and "creation science" appeared on the scene. The appearance of a small metal sphere in the heavens helped to kick-start the process.

The Sputnik Scare

In October 1957, the Soviet Union launched Sputnik, the world's first artificial satellite. The United States was shocked: the Communists had beaten the world's foremost democracy into space. How had this happened? As part of the soul-searching that took place after the Soviet triumph, the United States decided that the scientific establishment—including public school science instruction—was seriously in need of an overhaul. The newly established National Science Foundation beefed up funding for basic scientific research and also instituted directorates for education that would fund research to improve science education.

One goal was to improve the content and pedagogy of high school textbooks. It was an ambitious effort: scientists and master teachers were assembled to prepare textbooks in physics, chemistry, earth science, and biology. When university-level scientists began working with the NSF-funded Biological Sciences Curriculum Study (BSCS), they were shocked to discover the poor quality of extant textbooks. Evolution, the foundation of biology, was absent from almost all of them. They decided that in addition to improving the pedagogical approach to learning ("to get away from the 'parade-of-the-plant-and-animal-kingdoms' approach, to stress concepts and experimental science, and to encourage the personal involvement of students in their learning" [Moore 2002: 165]), the new BSCS books would treat evolution as it was treated in college-level texts: as an indispensable component of the biological sciences that students must understand in order to understand biology fully. In 1963, the first three BSCS textbooks were released, and all of them included evolution as a prominent theme (Grobman 1998).

The new BSCS approach brought about a revolution in textbooks. Partly because these new textbooks carried the stamp of approval of the National Science Foundation, but also because they were so much more interesting and up-to-date than extant books, school boards and textbook selection committees were eager to adopt them. Once the BSCS books began selling, commercial publishers began to try to produce books in the same mold (Skoog 1978: 24). As one of the BSCS writers described it:

> Subsequent events showed that nearly every objecting school board ended up adopting the books—evolution, sex, and all. Word was spreading the BSCS biology was the "new thing," and there were community pressures on school boards to be up to date, even if a little wicked, rather than behind the times and fully virtuous. Once this situation was understood, nearly every newly published biology book included an explicit discussion of evolution. (Moore 1976: 192–193)

After the decline of evolution in textbooks and science curricula after the Scopes trial, antievolutionists had not been very active; they had not needed to be. But the resurgence in the 1960s of evolution in textbooks generated new resistance to the teaching of evolution in the public schools. Although since the 1700s some supporters of a literal interpretation of the Bible had argued that scientific evidence existed to support their views, such arguments had diminished considerably after Darwin's *On the Origin of Species*. Now, in the mid-twentieth century, such views were being revived, partly in response to the increasing presence of evolution in textbooks and in the curriculum. Much as the Scopes trial in 1925 had been a response to the post-Darwinian appearance of evolution in the curriculum, so did the return of evolution to the curriculum in the 1960s spark a reaction from religious conservatives (Numbers 1992). But the Scopes-era reaction to evolution was almost entirely centered on religious objections: evolution should not be taught to children, they argued, because it was unbiblical and would lead children away from faith. By the mid-twentieth century, however, science was a far more powerful cultural force than it had been earlier, and antievolutionists sought to exploit its authority (Larson 2003).

If students were going to be taught evolution, antievolutionists argued, students also should be exposed to a biblical view. The frank advocacy of a religious view such as creationism in the public schools would of course be unconstitutional, but creationists reasoned that if creationism could be presented as an alternative *scientific* view—Creation Science—then it would deserve a place in the curriculum. No one was more important in shaping this approach than Dr. Henry M. Morris.

Creation Science and Henry M. Morris

The Genesis of Creation Science. Henry M. Morris is widely considered to be the father of the twentieth-century movement known as Creation Science. Morris is trained as a hydraulic engineer and began his career as a creationist with the publication in 1946 of his first book, *That You Might Believe*, while he was still in graduate school. The book and its successor, *The Bible and Modern Science* (1951), proclaimed a recent six-day (24 hours/day) creation, and a literal, historical Flood. These views were, of course, based on interpretations of Genesis, but the claim additionally was made that Special Creationism can be supported by the facts and theories of science. Although both of these books are still in print and continue to sell, the modern Creation Science movement crystallized in 1961 with the publication of Morris's book *The Genesis Flood*, written with the theologian John Whitcomb.

Like Morris's previous works, *The Genesis Flood* argued that most modern geological features could be explained by Noah's Flood, a view that had originally been popularized by the early twentieth-century Seventh-Day Adventist geologist George McCready Price (Numbers 1992). Termed Flood Geology, this view became the core of the new movement called Creation Science.

Morris provided the scientific references, and Whitcomb provided the theological arguments. The book's mix of theology and science is characteristic of Creation Science, and it continues to be widely read in Evangelical and Fundamentalist circles. *The Genesis Flood* proposed that there is scientific evidence that Earth is less than 10,000 years old, and that evolution was therefore impossible. This view became known as Young Earth Creationism. Fundamentalists were eager to claim scientific support for their religious views and use it to "balance" the teaching of evolution.

Morris worked tirelessly to strengthen the Evangelical antievolutionist movement. To promote scientific research supporting the young age of Earth and the universe, the special creation of all living things, and Noah's Flood, he worked with a group of conservative Christian scientists to found the Creation Research Society (CRS) in 1963, soon after the publication of *The Genesis Flood.* The *Creation Research Society Quarterly* (CRSQ) began publishing shortly thereafter, in 1964. Although in the early days the board included some non–Flood Geology proponents, the CRS eventually evolved into a Young Earth organization.

Unusually, for scientific societies, the CRS requires all voting members to sign a statement of belief. Reflecting the Young Earth Creationist orientation of Morris and other influential founders, the statement includes the following provisions (emphases in the original):

1. The *Bible* is the written Word of God, and because it is inspired throughout, all its assertions are historically and scientifically true in the original autographs. To the student of nature this means that the account of origins in *Genesis* is a factual presentation of simple historical truths.
2. All basic types of living things, including man, were made by direct creative acts of God during the Creation Week described in *Genesis*. Whatever biological changes have occurred since Creation Week have accomplished only changes within the original created kinds.
3. The great flood described in *Genesis*, commonly referred to as the Noachian Flood, was an historic event worldwide in its extent and effect.

Tenets of Creation Science. Special Creationism is a religious view accepted in whole or in part by many American Christians (see chapter 3). Creation Science reflects Special Creationism in that it professes that the universe came into being in its present form relatively suddenly, over a period of days rather than billions of years. The galaxies, Earth, and living things on Earth appeared during six 24-hour days of creation, according to this view. Creation Science, as outlined by Henry M. Morris, also proposes that the universe is young, with its age reckoned in thousands, rather than billions or even millions, of years. Creation Science includes these ideas derived from Special Creationism but adds that this account of creation can be supported with scientific data and theory: its proponents do not consider Creation Science to be limited to a religious view.

Creation Science argues that there are only two views, (Special) Creationism or evolution; thus arguments against evolution are arguments in favor of creationism. Literature supporting Creation Science thus centers on alleged examples of "evidence against evolution," which are considered not only disproofs of evolution but also positive evidence for creationism.

Creation Science Expands

To counter the BSCS and other evolution-based textbooks, in 1970 the CRS published its own high school–level biology textbook, *Biology: A Search for Order in Complexity*, which by its title revealed its orientation of seeking divine design in nature (Moore and Slusher 1970). The CRS textbook did not sell many copies, however, and ran into legal difficulty in several states because of its frankly religious orientation. In 1974, Henry Morris published a textbook of his own, *Scientific Creationism.* To try to avoid the criticisms that such books were Christian apologetics masquerading as science books, Morris published a "Christian schools edition," containing an extra chapter of biblical references, and a "general edition," which did not. Claims that religious references were not present in the general edition were not persuasive, however, when textbook selection committees encountered statements referring to such biblical events as the Flood of Noah and the Tower of Babel:

> The origin of civilization would be located somewhere in the Middle East, near the site of Mount Ararat (where historical tradition indicates the survivors of the antediluvian population emerged from the great cataclysm) or near Babylon (where tradition indicates the confusion of languages took place). (Morris 1974: 188)

The Institute for Creation Research. In 1972, Henry Morris and others founded the Institute for Creation Research (ICR) as the research division of the Bible-based Christian Heritage College, which Morris had founded two years earlier with the evangelist Timothy LaHaye. ICR became an independent institution in 1980, moving from the San Diego suburb of El Cajon to nearby Santee, California, where it currently resides in two large buildings. Now grown to an institution with a staff of over 40, the ICR began as and remains the flagship antievolution ministry.

The ICR has grown steadily since its inception in the early 1970s, taking pride in always ending the year in the black and never borrowing money for its building projects. To promote Creation Science, ICR conducts extensive outreach to churches and individuals. In any given week, ICR staff may be found around the country leading workshops, lecturing, or occasionally debating evolution with scientists. ICR's popular Back to Genesis program, begun in 1988, consists of two days of lectures, movies, and workshops for adults and children. Other programs, such as the Good Science Workshops, are aimed at school-age children, parents, and teachers, and focus on Creation Science education. ICR's foremost debater, Dr. Duane Gish, holds workshops on how to debate evolutionists. One of ICR's radio programs, *Science, Scripture and Salvation,* is aired on approximately 700 stations, and the other, a one-minute filler by director John Morris (Henry's son), "Back to Genesis," has 860 outlets (www.icr.org/radio/rad-hist.htm; accessed July 17, 2002).

Each month, ICR mails literature to 200,000 or more recipients. The monthly ICR mailing to supporters includes a newsletter, *Acts and Facts,* and one or more pamphlets, among them the semischolarly *Impacts,* and the much more evangelical *Back to Genesis* series. Often an advertising circular promoting books, videos, CDs, or other media is included, as well as a letter from the president of ICR and a request for financial support.

Early in ICR's history, Morris helped found Creation-Life Publishers, which through its Master Books division maintains an extensive catalog of antievolution books. Master Books are promoted not only by ICR but also by virtually every creationist organization large enough to sell merchandise. The catalog includes over 150 books.

The ICR also maintains the Museum of Creation and Earth History at its Santee headquarters, which was visited by an estimated 25,000 individuals during its first year (Anonymous 1993b). Remodeled in 1992, it currently reaches thousands of

schoolchildren each year, most of them homeschooled or attending Christian schools. Because of the religious orientation of the museum, few local public school teachers take their students to the ICR Museum. The museum presents a journey through the seven days of creation, mixing biblical with scientific references. True to Morris's concern with Flood Geology, there is a Noah's Ark diorama, presenting calculations of how many animals could have been housed on the Ark.

To promote the establishment of Creation Science, ICR supports a graduate school that offers master's degrees in science education, biology, geology, and astrogeophysics. The school is not accredited by the Western Association of Colleges and Schools, the accrediting agency for most other California institutions of higher learning, but by the TransNational Association of Christian Schools (TRACS). TRACS was founded by none other than Henry M. Morris, who served as its president for many years. TRACS accredits Bible-based institutions that pledge to promote Creation Science. In 1991 the federal Department of Education, over some protest, accepted TRACS as an accrediting agency for institutions of this type, which means, among other things, that students in such institutions can qualify for federal educational loans.

Most of the graduate school courses are taught during the summer. There are also annual trips down the Grand Canyon, where ICR geologist Steve Austin explains how the many layers of the canyon were formed by the receding waters of Noah's Flood.

Ken Ham and Answers in Genesis. In January 1987, a young Australian evangelist, Ken Ham, came to work for the ICR (Anonymous 1986). Ham had founded the Australian Creation Science Foundation in 1978 and had built it into a successful Young Earth ministry. In April 1988, ICR announced the institution of the Back to Genesis program, consisting of two-and-a-half-day public meetings built around Creation Science, but being more explicitly evangelical and religious than other ICR programs (Anonymous 1988a, 1988b). Although other staff members were necessarily involved, the Back to Genesis program relied heavily upon Ham. The more evangelical focus of its meetings may be due to the fact that Ham lacks a background in science, unlike most other ICR professional staff. By all accounts, the former teacher was a popular and successful evangelist, and the Back to Genesis programs began to play a larger role in ICR activities.

Ham also wrote the new *Back to Genesis* evangelical pamphlets that, beginning in January 1989, accompanied the ICR newsletter, *Acts and Facts*, and Ham-led Back to Genesis revivals soon were held at least once a month somewhere in the United States. By August 1993, the Back to Genesis program had apparently expanded to the limits of ICR staff capabilities, and an article appeared in *Acts and Facts* encouraging churches to sign up for other, smaller ICR programs such as the Case for Creation seminars, "as well as speakers for pulpit supply, parent/teacher science workshops, school assemblies, conventions, campus conferences, and other types of meetings, even field trips are possible, especially with graduate student [*sic*], wherever a creationist message is in demand and can be scheduled. Fees are very reasonable compared to those of other types of speciality speakers" (Anonymous 1993a: 5).

In January 1994, Ham moved to adjunct faculty status at the ICR's graduate school and left for Florence, Kentucky, to establish a branch of the Australian Creation

Science Foundation. Ham's Creation Science Ministries changed its name to Answers in Genesis (AIG), and a few years later (in November 1997) the Australian Creation Science Foundation also became Answers in Genesis. The Australian and U.S. organizations formed the core of an international movement with branches in the United States, Australia, Canada, New Zealand, and Great Britain. All members receive as a membership premium the glossy Australian creationist magazine *Creation ex Nihilo*.

Ham's organization has become the second-largest YEC organization in the nation. According to Internal Revenue Service records (available for all nonprofit organizations), AIG's income in 2000 was in excess of $5 million. AIG sponsors lectures and workshops led by Ham and other employees, and his monthly newsletter, *Answers Update*, offers books, videos, audiotapes, and other resources promoting YEC.

Other Young Earth Creation Ministries. In addition to the ICR, other national Young Earth Creationist organizations have heeded the Creation Science message of Henry Morris. The Bible-Science Association (B-SA), founded in 1964, focused on spreading the message of Creation Science to the general public rather than on publishing scientific research. There were, in fact, some tensions between the B-SA and the CRS. The B-SA published the *Bible Science Newsletter* until the late 1990s, and then, falling upon lean times, cut back publication and activities to focus on a creationist radio program (first aired in 1987) named *Creation Moments*. The B-SA itself became Creation Moments, Inc., in 1997, and concentrates wholly on producing short radio programs.

Creation Science is communicated to the public primarily through ICR publications and also the less widely distributed *Creation Research Society Quarterly*, published by the CRS. In 1978, Students for Origins Research began publishing the newsletter *Origins Research* from Santa Barbara, California. Although they are more moderate than the B-SA, the students usually promoted a Young Earth orientation. In 1996, *Origins Research* changed title and format and emerged as *Origins and Design*, and Students for Origins Research morphed into Access Research Network, shedding its overt Young Earthism to promote Intelligent Design Theory (see chapter 6).

There also are regional and local organizations that promote the Young Earth creationist views of Henry Morris, such as the Paulden, Arizona–based Van Andel Research Center (named after Jay Van Andel, founder of the Amway company), headed by an adjunct professor of biology at ICR, John R. Meyer. In Grand Junction, Colorado, Dave and Mary Jo Nutting (ICR graduate school alumni) operate the Alpha Omega Institute, which provides creationist geology and natural history tours of the Rockies and surrounding areas, and school assemblies for Christian and public schools. A St. Louis, Missouri–based organization, the Creation-Science Association of Mid-America, was instrumental in the late 1990s in providing information to Kansas board of education members wishing to promote Creation Science. The Pittsburgh-based Creation Science Fellowship, Inc., has been promoting Young Earth Creationism since 1980 and sponsors periodic international conferences on creationism. Several independent evangelists focus on creationism, including Kent Hovind of Pensacola, Florida, Carl Baugh of Glen Rose, Texas, and Walter Brown of the Center for Scientific Creation in Phoenix, Arizona. Some national televangelists such as D. James Kennedy (Coral Ridge Ministries, Florida), Hank Hanegraaff (Rancho Santa

Margarita, California), and John Ankerberg (Ankerberg Theological Research Institute, Chattanooga, Tennessee) regularly present programs criticizing evolution and rely on and promote creationist views. The nationwide Maranatha Campus Ministries also promotes Creation Science. A sizable corpus of antievolutionary material consisting of books, videos, compact discs, filmstrips, tape recordings, posters, and curricula promoting Young Earth Creationism is publicly available from these organizations and individuals and their Web sites.

Equal Time for Creation and Evolution

Undoing Scopes. By the early 1960s, evolution was returning to science textbooks and classrooms after largely having been absent since the 1930s. It is not coincidental that Whitcomb and Morris's *The Genesis Flood* was published in 1961 and the ICR was founded a few years later: the increased exposure of public school students to evolution was a cause for alarm among conservative Christians. In Arizona, opposition to the use of BSCS books in the Phoenix school district stimulated state legislators to introduce legislation that would require "equal time and emphasis to the presentation of the doctrine of divine creation, where such schools conduct a course which teaches the theory of evolution" (the legislation did not pass) (Larson 2003: 97). But the renewed textbook emphasis on evolution generated conflicts in states with antievolution laws for teachers who wished to teach modern science but who would thereby be breaking the law.

In 1965, Arkansas was one of the few remaining states with Scopes-era antievolution laws still on the books (the others were Tennessee, Louisiana, and Mississippi [Larson 1997]). In that year, the Arkansas Education Association (AEA) decided to challenge the state's antievolution law, partly because the presence of evolution in textbooks put teachers on a collision course with the law. Rather than a Scopes-style teacher defendant who would be prosecuted for breaking the law, the AEA instead challenged the law itself with a teacher plaintiff who sought to legally teach evolution (Moore, 1998). Arkansas teacher Susan Epperson argued that the Arkansas antievolution law was unconstitutional because it violated her freedom of speech, and a coplaintiff, a father of a student, argued for the right of the student to learn the banned subject. The trial itself was very short, taking only about two hours; the judge ruled that the antievolution law was unconstitutional. To the surprise of Epperson and the AEA, the Arkansas Supreme Court reversed the lower court in a two-sentence decision in 1967.

The case was appealed to the U.S. Supreme Court, which ruled in 1968 in *Epperson v. Arkansas* that the antievolution law was unconstitutional because it "selects from the body of knowledge a particular segment which it proscribes for the sole reason that it is deemed to conflict with a particular religious doctrine." The First Amendment requires schools to be neutral toward religion; to ban a subject (evolution) because a religious view (Fundamentalism) finds it objectionable violates the Establishment Clause of the First Amendment. Finally, in 1968, 43 years after the Scopes trial, it was unlawful to ban the teaching of evolution.

Epperson had more of a psychological effect on antievolutionism than an actual one, as the Arkansas and other antievolution laws had hardly ever been enforced

(Larson 2003). But if evolution could not be banned, how could children be protected from it? Keeping evolution out of the classroom was obviously not possible, with evolution widely being included in textbooks. Teaching the Bible along with evolution was one solution.

"Neither Advances nor Inhibits Religion." In 1963, the Supreme Court struck down laws requiring prayer in public schools (*Abington School District v. Schempp*). The First Amendment of the U.S. Constitution sets forth freedoms of religion, speech, and assembly. The Religion Clause reads, "Congress shall make no law respecting the establishment of religion, nor inhibiting the free exercise thereof." The Establishment Clause prohibits the state from promoting religion, and the Free Exercise Clause prohibits the state from inhibiting or restricting religion. In *Schempp*, the justices clearly stated the requirement for religious neutrality in the public schools, stating that "to withstand the strictures of the Establishment Clause there must be a secular legislative purpose and a primary effect that neither advances nor inhibits religion."

William Jennings Bryan had argued that neutrality consisted of teaching *neither* evolution nor creationism in the schools: antievolution laws removed evolution from the curriculum so that students would not be exposed to what some considered an antireligious doctrine. As evolution returned to textbooks and to the curriculum, creationists protested that the classroom was no longer neutral. To restore neutrality, they argued, *both* evolution and creationism should be taught. Even before the *Epperson* decision struck down antievolution laws, parents Nell Segraves and Jean Sumrall petitioned the California Board of Education in 1963 to restore neutrality to the classroom by adding creationism to the curriculum if evolution were taught.

A *Movement Builds.* Henry Morris's original approach to promoting Creation Science was to reach out to the scientific and educational communities: Morris conceived of the ICR as a research and educational institute that eventually would persuade the academic community of the value of Creation Science. He believed that after scientists, educators, and the public understood Creation Science, the subject would trickle down to the school science curriculum. Other creationist organizations (such as Nell and Kelly Segraves's Creation Science Research Center) sought to promote creationism through political action; Morris, perhaps because of his background as a former college professor, preferred to work through education (Numbers 1992).

Scientists and educators, however, ignored Creation Science. The top-down approach wasn't working, so ICR shifted its strategy toward the grass roots. Although never embracing the Creation Science Research Center's approach of filing lawsuits to force the teaching of creationism, the ICR nonetheless encouraged citizens to take an active role in promoting Creation Science at the local level. In ICR's publication *Impacts*, lawyer Wendell Bird encouraged local citizens to present school boards with "resolutions" encouraging the teaching of Creation Science in science curricula (Bird 1979).

The model resolution laid out the definition of Creation Science: "special creation from a strictly scientific standpoint is hereinafter referred to as 'scientific creationism'" (Bird 1979: ii). It claimed that the presentation of only evolution in the classroom "without any alternative theory of origins" was unconstitutional "because it undermines their [students'] religious convictions," it would require students to attest to course materials they did not believe in, and it "hinders religious training by parents" (Bird 1979: ii). The resolution claimed that evolution-only teaching would promote belief systems such as "religious Liberalism, Humanism, and other religious faiths." It claimed that the "theory of special creation is an alternative model of origins at least as satisfactory as the theory of evolution and that the theory of special creation can be presented from a strictly scientific standpoint without reference to religious doctrine" (Bird 1979: ii). School districts were then urged to give "balanced treatment to the theory of scientific creationism and the theory of evolution" in all aspects of the curriculum, including classroom time, textbook contents, and library materials.

Even before the ICR model resolution appeared, a conservative Christian layman, Paul Ellwanger, had submitted his own resolution to the Anderson, South Carolina, school district, proposing a "balanced treatment of evolution and creation in all courses and library materials dealing in any way with the subject of origins" (Anonymous 1979: ii). Feedback between the Ellwanger and ICR resolutions resulted in Ellwanger preparing sample legislation for districts or states to pass.

The legislation presented two alternative—and allegedly scientifically equivalent—views of "origins": "evolution science" and "creation science," both of which should be taught to maintain a "balanced" curriculum. If evolution were taught, schools would be required also to teach Creation Science. Inspired by Ellwanger's efforts, a movement began to introduce his bills in state legislatures. The late 1970s campaign to promote equal time for Creation Science legislation was truly a grassroots effort, in the classic American tradition. The campaign spread largely by word of mouth and did not yet have the blessing or resources of national religious denominations or religiously oriented political organizations such as the Moral Majority. Although legislators in Ellwanger's home state of South Carolina failed to pass an Ellwanger bill, legislation soon began appearing in other states.

By the early 1980s, equal time legislation had been introduced in at least 27 states, including Alabama, Colorado, Florida, Illinois, Indiana, Iowa, Louisiana, Missouri, Nebraska, Oklahoma, Oregon, South Carolina, Texas, and Washington (Moyer 1981: 2), Georgia, Kentucky, Minnesota, New York, Ohio, Tennessee, West Virginia, and Wisconsin (Anonymous 1981: back cover), Maryland (Weinberg 1981b: 1), Arkansas (Weinberg 1981a: 1), Mississippi, Arizona, and Kansas (Weinberg 1982: 1). All died in committee, except for those in Arkansas and Louisiana. Many scientists and educators were involved in campaigns to prevent the passage of equal time legislation. Creation Science finally was receiving attention from scientists, though not the kind Henry M. Morris had desired.

McLean v. Arkansas. That Arkansas in 1981 was the first state to pass a creation and evolution equal time bill has an ironic twist: as discussed earlier in this chapter, in 1968, Arkansas had been the site of the Supreme Court case that struck down Scopes-era antievolution laws. Now it was to be in the spotlight again as the site of the first challenge to "equal time" legislation.

Arkansas Act 590 proposed "balanced treatment" for "evolution-science" and "creation-science," defining Creation Science as:

(1) Sudden creation of the universe, energy, and life from nothing;
(2) The insufficiency of mutation and natural selection in bringing about development of all living kinds from a single organism;
(3) Changes only within fixed limits of originally created kinds of plants and animals;
(4) Separate ancestry for man and apes;
(5) Explanation of the earth's geology by catastrophism, including the occurrence of a worldwide flood; and
(6) A relatively recent inception of the earth and living kinds.

"Evolution science" was defined as:

(1) Emergence by naturalistic processes of the universe from disordered matter and emergence of life from nonlife;
(2) The sufficiency of mutation and natural selection in bringing about the development of present living kinds from simple earlier kinds;
(3) Emergence by mutation and natural selection of present living kinds from simple earlier kinds;
(4) Emergence of man from a common ancestor with apes;
(5) Explanation of the earth's geology and the evolutionary sequence by uniformitarianism; and
(6) An inception several billion years ago of the earth and somewhat later of life.

According to Act 590, presenting only evolution in the schools created a hostile climate for religious students, undermining "their religious convictions and moral or philosophical values," and violating "protections of freedom of religious exercise and of freedom of belief and speech for students and parents" (Anonymous 1983: 18). Teaching only evolution was held to be a violation of academic freedom, "because it denies students a choice between scientific models and instead indoctrinates them in evolution-science alone" (Anonymous 1983: 18). Creation Science was presented as a "strictly scientific" view.

Upon passage, Governor Frank White signed Act 590, and the bill was immediately challenged by the Arkansas American Civil Liberties Union (AACLU). Plaintiffs in the lawsuit included religious leaders, science education organizations, civil liberty organizations, and several individual parents. Methodist clergyman William McLean was the lead plaintiff, and he was joined by the bishops or other spokespersons for the Arkansas Episcopal Church, the United Methodists, Roman Catholics, African Methodist Episcopalians, Presbyterians, and Southern Baptists. Also joining the suit were the Arkansas Education Association, the National Association of Biology Teachers, the American Jewish Congress, the Union of American Hebrew Congregations, the American Jewish Committee, and the National Coalition for Public Education and Religious Liberty. The presence of so many religious plaintiffs helped to defuse the argument that opposition to the bill equated to opposition to religion.

McLean v. Arkansas was tried in Federal District Court. The AACLU received considerable assistance from a large New York law firm, Skadden, Arps, Slate, Meagher and Flom, which offered its services pro bono. The AACLU would argue that because Creation Science was inherently a religious idea, its advocacy as required by Act 590 would violate the Establishment Clause. Furthermore, because Creation

Science was not scientific, there was no secular purpose for its teaching (Herlihy 1983). The state, defending the law, had to argue the opposite: that Creation Science was scientific, and thus its advocacy would have a secular purpose. The state ignored the issue of whether Creation Science was religious. Each side brought in witnesses to testify in favor of its position. Much time was spent in the trial over the definition of science and whether Creation Science fulfilled it.

The state was reluctant to put Henry Morris or any other ICR spokesperson on the stand, notwithstanding the prominence of these people in the creationism movement (Larson 2003: 162). Because of the Christian apologetic nature of so much of Morris's writings, the defense was unwilling to have Morris cross-examined: it would be apparent that Creation Science was primarily a religious view, and the state's case would be lost from the beginning. Witnesses for the defense, therefore, consisted of other, less well-known creationists and some noncreationist scientists who questioned some aspects of evolution. An example of the latter was the British astrophysicist Chandra Wickramasinghe, who argued against the standard chemical origin-of-life model. His explanation for the origin of life on Earth was not Special Creation, however, but a natural explanation, having to do with the seeding of life on Earth with organic molecules from comets. When questioned about Creation Science, he stated that "no rational scientist could believe that the earth was less than one million years old" (Holtzman and Klasfeld 1983: 95). In the aftermath of the case, creationists blamed the state attorney general for not using the "strongest advocates" of Creation Science, although such a move would have been legal suicide.

The plaintiffs assembled a cast of eminent scholars—scientists, theologians, philosophers of science, sociologists, and educators—to make the case that Creation Science was not science but a form of sectarian religion. Included were three members of the National Academy of Sciences, the nation's most prestigious scientific organization. Press accounts attest to the articulateness and depth of knowledge of these witnesses and to the superiority of the Skadden, Arps lawyers in cross-examination and in the general presentation of the case. The consensus was that the defense was simply outgunned. It was so apparent to the plaintiffs that their case would be successful that lawyers and witnesses had their victory party on the third day of the trial (Ruse 1984: 338).

Indeed, when the judge issued his decision, it was in favor of the plaintiffs; Act 590 was declared unconstitutional. In a strongly worded decision, Judge William Overton relied upon a 1971 Supreme Court decision, *Lemon v. Kurtzman*, which had established three tests for determining whether a law or practice violated the Establishment Clause. The three "prongs" of Lemon, as they have come to be called, are the "purpose," "effect," and "entanglement" rules.

Lemon (as did the earlier *Schempp*) requires that the "statute must have a secular legislative purpose"; if the legislature's purpose in passing the law was to advance religion, then the law fails (403 U.S. at 612–613). Judge Overton ruled that the legislative history of the law clearly demonstrated that the legislators intended to promote a religious view.

"[S]econd, its principal or primary effect must be one that neither advances nor inhibits religion" (403 U.S. at 612). This "effect" prong was likewise judged to be violated by Act 590; Judge Overton decided that requiring Creation Science to be taught would promote a sectarian religious view, because Creation Science was a religious view, not a science. Much of the legal decision, in fact, was devoted to showing how Creation Science did not meet a general definition of science accepted by practitioners.

Lemon also states that ". . . the activity must not foster 'an excessive government entanglement with religion'" (403 U.S. at 613). Judge Overton ruled that because the classroom must not be a place for religious proselytization, the administration would have to monitor teachers and instructional material to guard against willing or unwitting advancement of religion.

Because the *McLean* decision declared so strongly that Act 590 was unconstitutional, the state declined to appeal the case to the Court of Appeals. Equal time for Creation Science and evolution had failed in Arkansas, but a law very similar to the Arkansas law had been introduced into neighboring Louisiana only a few months before the *McLean* decision.

The Louisiana Equal Time Law. The law passed by the Louisiana legislature in the spring of 1982 was another Ellwanger clone, with a few modifications intended to make it more likely to pass constitutional muster. The plaintiffs in *McLean* had been able to show that Act 590's definition of Creation Science paralleled biblical literalist creationism, a similarity that figured into Judge Overton's decision to strike it down. The framers of the Louisiana law, "Balanced Treatment for Creation-Science and Evolution-Science in Public School Instruction," sought to mount a stronger case by not defining Creation Science in recognizably religious terms. Again the ACLU challenged the law in Federal District Court, but because proponents of the law also requested an injunction, courts had to sort out jurisdictional issues, and both cases slogged through the courts for several years. Finally the case was heard by the Federal District Court. Rather than holding a full trial, as in Arkansas, the District Court tried the case by "summary judgment": the judge accepted written statements from both sides and decided the outcome of the case based on these documents.

In 1985, the Federal District Court decided that the law was unconstitutional because it advanced a religious view by prohibiting the teaching of evolution unless creationism—a religious view—was also taught. The Court of Appeals agreed, and finally the case made its way to the Supreme Court in 1987. The highest court concurred with the lower courts as follows:

> The preeminent purpose of the Louisiana Legislature was clearly to advance the religious viewpoint that a supernatural being created humankind. . . . The Louisiana Creationism Act advances a religious doctrine by requiring either the banishment of the theory of evolution from public school classrooms or the presentation of a religious viewpoint that rejects evolution in its entirety. (*Edwards v. Aguillard*, 482 U.S. 578 [1987])

"Equal time for Creation Science" was no longer a legal option in the schools of the United States. Shortly after the filing of the *Edwards* decision, however, creationist attorney Wendell Bird wrote an *Impact* pamphlet for the ICR in which he proposed the next strategy of antievolutionism: the repackaging of Creation Science so that it might survive such Establishment Clause challenges as had doomed it in Arkansas and Louisiana. The next, Neocreationist, stage of American antievolutionism was beginning to evolve.

REFERENCES CITED

Anonymous. 1979. Creation Science and the Local School District. *ICR Impact*, January: 4.

Anonymous. 1981. Update on Creation Bills and Resolutions. *Creation/Evolution* 2 (1): 1–44.

Anonymous. 1983. Act 590 of 1981. In *Creationism, Science, and the Law*, edited by M. C. La Follette. Cambridge, MA: MIT Press.

Anonymous. 1986. Ken Ham Joins ICR Staff. *Acts and Facts*, December: 1–4.

Anonymous. 1988a. "Back to Genesis" Programs Have a Successful Start. *Answers in Genesis*, April: 1–4.

Anonymous. 1988b. First "Back-to-Genesis" Program a Resounding Success; Over 2000 in New England States Hear Creation Message. *Acts and Facts*, July: 1, 5.

Anonymous. 1990. *The World's Most Famous Court Trial: Tennessee Evolution Case*, 2nd ed. Dayton, TN: Bryan College.

Anonymous. 1993a. Creationist Speakers Available. *Acts and Facts*, August: 5.

Anonymous. 1993b. ICR Museum Visited by 25,000 in First Year. *Acts and Facts*, November: 1.

Armstrong, Karen. 2000. *The Battle for God: A History of Fundamentalism*. New York: Ballantine Books.

Bird, Wendell R. 1979. Resolution for Balanced Presentation of Evolution and Scientific Creationism. *ICR Impact*, May: 4.

Carnegie, Andrew. 1889. Wealth. *North American Review*, June: 653–664.

Grabiner, Judith V., and Peter D. Miller. 1974. Effects of the Scopes Trial. *Science* 185 (4154): 832–837.

Grobman, Arnold B. 1998. National Standards. *American Biology Teacher*, October: 562.

Herlihy, Mark E. 1983. Trying Creation: Scientific Disputes and Legal Strategies. In *Creationism, Science and the Law: The Arkansas Case*, edited by M. C. La Follette. Cambridge, MA: MIT Press.

Holmes, S. J. 1927. Proposed Laws Against the Teaching of Evolution. *Bulletin of the American Association of University Professors* 13 (8): 549–554.

Holtzman, Eric, and David Klasfeld. 1983. The Arkansas Creationism Trial: An Overview of the Legal and Scientific Issues. In *Creationism, Science, and the Law: The Arkansas Case*, edited by M. C. La Follette. Cambridge, MA: MIT Press.

Larson, Edward J. 1997. *Summer for the Gods: The Scopes Trial and America's Continuing Debate over Science and Religion*. New York: Basic Books.

Larson, Edward J. 2003. *Trial and Error: The American Controversy over Creation and Evolution*, third ed. New York: Oxford University Press.

Marsden, George M. 1980. *Fundamentalism and American Culture: The Shaping of Twentieth-Century Evangelicalism, 1870–1925*. New York: Oxford University Press.

Moore, John A. 1976. Creationism in California. In *Science and Its Public: The Changing Relationship*, edited by Gerald Horton and William A. Blanpied. Dordrecht, Netherlands: Reidel.

Moore, John A. 2002. *From Genesis to Genetics: The Case of Evolution and Creationism*. Berkeley: University of California Press.

Moore, John N., and Harold S. Slusher, eds. 1974. *Biology: A Search for Order in Complexity*, rev. ed. Grand Rapids, MI: Zondervan.

Moore, Randy. 1998. Thanking Susan Epperson. *American Biology Teacher*, November/December: 642–646.

Morris, Henry M., ed. 1974. *Scientific Creationism*. San Diego: Creation-Life Publishers.

Moyer, Wayne. 1981. Legislative Initiatives. *Scientific Integrity*, June: 1–4.

Numbers, Ronald. 1992. *The Creationists*. New York: Knopf.
Ruse, Michael. 1984. A Philosopher's Day in Court. In *Science and Creationism*, edited by A. Montagu. New York: Oxford University Press.
Skoog, Gerald. 1978. Does Creationism Belong in the Biology Curriculum? *American Biology Teacher* 40 (1): 23–29.
Skoog, Gerald. 1979. The Topic of Evolution in Secondary School Biology Textbooks: 1900–1979. *Science Education* 63: 621–640.
Weinberg, Stanley. 1981a. Memorandum to Liaisons for Committees of Correspondence. 1 (April): 1.
Weinberg, Stanley. 1981b. Memorandum to Liaisons for Committees of Correspondence. 1 (November): 1–4.
Weinberg, Stanley. 1982. Memorandum to Liaisons for Committees of Correspondence. 2 (February): 1–4.

CHAPTER 6

Neocreationism

In 1987, the Supreme Court decision *Edwards v. Aguillard* struck down a Louisiana law requiring "equal time" for creationism and evolution. Creationism is a religious idea, said the Court, and the First Amendment prohibits government from promoting religion. Antievolution strategies subsequently were developed that avoided the use of any form of the words "creation," "creator," or "creationism." In effect, proponents shifted their strategy from proposing to balance evolution with Creation Science to proposing to balance evolution with the teaching of Creation Science–like "scientific alternatives to evolution" or "evidence against evolution"—avoiding referring to such purported disciplines as "creationism." A school district in Louisville, Ohio, which had an equal time regulation in place before *Edwards*, rewrote the science curriculum to "avoid mention of creationism in its curriculum guide, calling it alternative theories to evolution and adding it to the science classes" (Kennedy 1992). The avoidance of Creation Science terminology and the development of Creation Science–like "alternatives to evolution," plus the repackaging of Creation Science content into "evidence against evolution," constitute what I have called neocreationism, which continues into the twenty-first century. These approaches were encouraged by creationist interpretations of the *Edwards v. Aguillard* decision.

THE EDWARDS DECISION AND NEOCREATIONISM

Justice William Brennan wrote the Supreme Court decision striking down Louisiana's "balanced treatment" law. Seven justices signed the decision; Justice Antonin Scalia wrote a dissent joined by Chief Justice William Rehnquist. *Edwards v. Aguillard* was argued more narrowly than the earlier *McLean v. Arkansas* decision, which, in contrast, had struck down Arkansas's equal time law after finding that it violated all three prongs of *Lemon*. *Edwards* declared that the Louisiana equal time law violated the first prong (purpose) of *Lemon* and did not extend the argument to the other prongs (as had *McLean*). Another difference between the two cases appeared in the treatment of Creation Science as science: *McLean*—after a full trial—declared that Creation Science failed the test; *Edwards*—decided on a summary judgment—did not take a stand on whether Creation Science qualified as science. Instead, it firmly identified Creation Science as promoting the religious idea of a Creator, making it unconstitutional to advocate in the public schools.

> The Act impermissibly endorses religion by advancing the religious belief that a supernatural being created humankind. The legislative history demonstrates that the term "creation science," as contemplated by the state legislature, embraces this religious teaching. The Act's primary purpose was to change the public school science curriculum to provide persuasive advantage to a particular religious doctrine that rejects the factual basis of evolution in its entirety. Thus, the Act is designed either to promote the theory of creation science that embodies a particular religious tenet or to prohibit the teaching of a scientific theory disfavored by certain religious sects. In either case, the Act violates the First Amendment. (*Edwards v. Aguillard*, p. 579)

The *Edwards* decision, however, suggested loopholes that could be, and were, seized upon by creationists. One was a statement recognizing the extant ability of teachers to "supplant the present science curriculum with the presentation of theories, besides evolution, about the origin of life." Such theories, however, had to be secular and not religious:

> We do not imply that a legislature could never require that scientific critiques of prevailing scientific theories be taught. . . . In a similar way, teaching a variety of scientific theories about the origins of humankind to schoolchildren might be validly done with the clear secular intent of enhancing the effectiveness of science instruction. (*Edwards v. Aguillard*, pp. 593–594)

This wording encouraged antievolutionists to argue for the teaching of "scientific alternatives to evolution."

"SCIENTIFIC ALTERNATIVES TO EVOLUTION": ABRUPT APPEARANCE THEORY

Creation Science is of course the original "scientific alternative to evolution," but it had been identified as a religious view in *McLean* and could not constitutionally be advocated in the public schools. Lawyer Wendell Bird, who had advised the Institute for Creation Research on legal matters and who had in fact argued Louisiana's position before the Supreme Court in the *Edwards* case, proposed a new "scientific alternative" to evolution that he claimed was distinct from Creation Science. His view, which he dubbed "Abrupt Appearance Theory," was, however, indistinguishable in content from Creation Science.

While a graduate student at Yale University in the mid-1970s, Bird had written an article for the *Yale Law Journal* arguing for the constitutionality of teaching Creation Science. It was this article and Bird's later work as staff attorney for ICR that shaped the argument that Creation Science was a legal alternative to evolution, which, as a supposedly purely scientific position, could be taught without violating the Establishment Clause. Although this argument was unpersuasive to judges in both the *McLean* and *Edwards* cases, both the District and Supreme Courts recognized that it is indeed legal to teach a secular, nonreligious, truly scientific alternative to evolution. Although neither courts nor scientists have recognized such an alternative, Bird's Abrupt Appearance Theory was the first public post-*Edwards* attempt at formulating such an alternative.

The phrase "abrupt appearance" was part of the definition of Creation Science in literature presented by the creationist side in the *Edwards v. Aguillard* case. Creation Science in fact was defined in *Edwards* as including "the scientific evidences for creation and inferences from those scientific evidences," but also including "origin through abrupt appearance in complex form." Bird reworked his brief for the *Edwards* case into *The Origin of Species Revisited*, published in 1987. Abrupt Appearance Theory was held to be the scientific evidence for the sudden appearance of all living things—in fact, the entire universe—in essentially its present form. No material or transcendent agent was identified as causing this event; Bird was meticulous in avoiding any references that could be interpreted as religious and would therefore expose Abrupt Appearance Theory to the same First Amendment challenges as Creation Science.

Consciously attempting to distance himself from religious creationism, Bird identified two scientific alternatives to explain "origins": evolution and Abrupt Appearance Theory. Evolution was defined broadly as encompassing cosmological (stellar) evolution, biochemical evolution (the origin of life), and biological evolution (the common ancestry of living things) (Bird 1987: 17). Abrupt Appearance Theory contrasts sharply with the continuous unfolding of the universe expressed in evolutionary theory: "The theory of abrupt appearance involves the scientific evidence that natural groups of plants and animals appeared abruptly but discontinuously in complex form, and also that the first life and the universe appeared abruptly but discontinuously in complex form" (Bird 1987: 13).

The essence of Abrupt Appearance Theory, therefore, is discontinuity: stars and galaxies appear abruptly, and life and groups of living things appear abruptly, much as in the religious view of Special Creation. Abrupt Appearance Theory thus encompasses Creation Science and other religious views—though it is claimed to have a "totally empirical basis" (Bird 1987: 13).

> This theory of abrupt appearance is different from the theories of creation, vitalism, panspermia, and similar concepts. Discontinuous abrupt appearance is a more general theory and a more scientific approach than scientific views of creation, vitalism, or panspermia, although they can be formulated as submodels of abrupt appearance. (Bird 1987: 20)

Although mammoth in its scope (its two volumes purport to summarize scientific, pedagogical, philosophical, and legal aspects of the creation/evolution debate) and prodigious in the number of citations from both the scientific and creationist literature, *The Origin of Species Revisited* is rarely cited today in creationist literature. It was, and remains, ignored in the scientific literature, and after the mid-1990s virtually disappeared from the political realm as well. It has been supplanted by another "alternative to evolution" that was evolving parallel to it.

"SCIENTIFIC ALTERNATIVES TO EVOLUTION": INTELLIGENT DESIGN

"Intelligent Design" (ID) refers to a movement that began a few years before the *Edwards* decision and solidified in the few years after it. Like Creation Science, ID is presented as a "scientific alternative to evolution," but it has far fewer obviously religious references and has been more successful in appealing to Christians who are not biblical literalists.

The Origin of Intelligent Design

Intelligent Design creationism can be dated from the publication of *The Mystery of Life's Origin* (Thaxton et al. 1984). The three authors proposed that the origin of life not only currently was unexplained through natural causes, but also *could not* be explained through natural causes. As the biologist Dean H. Kenyon wrote in the introduction, "It is fundamentally implausible that unassisted matter and energy organized themselves into living systems" (Thaxton et al. 1984: viii). The essential scientific claim of ID was made clear from the very beginning: some things in biology are categorically unexplainable through natural causes.

Encouragement for *The Mystery of Life's Origin* came from Jon Buell, a former campus minister who became president of the Dallas-based conservative Christian organization the Foundation for Thought and Ethics (FTE). He recruited the historian and chemist Charles Thaxton, the engineer Walter Bradley, and the geochemist Roger Olsen to write a document on scientific difficulties concerning the origin of life, which became *The Mystery of Life's Origin*. Buell, Thaxton, Bradley, Olsen, and others associated with the FTE proposed a new form of creationism that did not rely directly on the Bible: there were no references to a universal Flood, to the special creation of Adam and Eve or any other creature, or to a young Earth. But paralleling Creation Science, *Mystery* emphasized supposed scientific problems of evolution. *Mystery* mostly stuck to science, with only brief references in an epilogue to the necessity of "intelligence" being involved in the origin of life. Much as had Bird in proposing Abrupt Appearance Theory, the authors were agnostic on the identity of the creative agent. They offered the suggestion of Hoyle and Wickramasinghe (1979) that life on Earth was produced by extraterrestrials of high intelligence, although the authors expressed their preference for creation by God.

The next ID product to emerge was again from FTE: the 1989 high school biology supplementary textbook, *Of Pandas and People*, written by the biologists Percival Davis and Dean Kenyon. Originally titled *Biology and Origins*, the book was submitted to secular publishers for over two years before one was found—a small Texas press that specialized in agricultural materials (Scott 1989). By this time, the nascent movement had settled upon "Intelligent Design" as its sobriquet, and this term appeared in *Pandas*.

Although *Pandas* soon was proposed for adoption as an approved (and thus purchasable using state funds) textbook in at least two states (Idaho and Alabama) and in several school districts, its supporters were unsuccessful. The ID movement remained largely unnoticed until the publication in 1991 of *Darwin on Trial* by University of California–Berkeley law professor Phillip Johnson.

Because previously the antievolution movement had been based in small, nonacademic, nonprofit organizations such as ICR and FTE, the publication of an antievolution book by a tenured professor at a major secular university came as a surprise to the educated public. Although books by Henry Morris are ignored by the scientific community, a few scientists reviewed *Darwin on Trial* in popular publications such as *Scientific American*, and discussions of this new form of antievolutionism appeared in the popular press. Scientists uniformly criticized what they considered to be uninformed science in Johnson's book. On the other hand, educated conservative Christians, for whom Creation Science was unacceptable because of its often outlandish scientific claims, found Johnson's message very attractive indeed. Largely because of its more respectable pedigree, Intelligent Design obtained considerably more coverage in the popular media than Creation Science—though the latter boasts many more organizations and activists than ID.

What, in detail, are the claims of ID? As is the case with Creation Science, ID combines a scholarly focus with an effort to promote a sectarian religious view.

The Scholarly Focus of Intelligent Design

Intelligent Design proponents posit that the universe, or at least components of it, have been designed by an "intelligence." They also claim that they can empirically distinguish intelligent design from that produced through natural processes (such as natural selection). This is done through the application of two complementary ideas, one promoted by a biochemist and the other by a philosopher/mathematician.

Irreducible Complexity. The biochemist Michael Behe contends that intelligence is required to produce irreducibly complex cellular structures (ones which couldn't function if a single part were removed) because such structures *could* not be produced by the incremental additions of natural selection (Behe 1996).

Critics of Behe have pointed out that it is not clear that irreducibly complex structures actually exist—except perhaps by definition. Critics have also argued that the examples Behe gives of irreducibly complex structures can often be reduced and still be functional. Behe commonly uses a mousetrap as his example of an irreducibly complex structure, claiming that if any one of the five basic parts of a mousetrap (platform, hammer, spring, catch, and hold-down bar) is removed, it can no longer catch mice. (See Figure 6.1.) Scientists gleefully set about producing four-part, three-part, two-part, and even one-part mousetraps to demonstrate the reducibility of Behe's prime example of an irreducibly complex structure.

Similarly, supposedly irreducibly complex biochemical structures such as the bacterial flagellum can function with fewer parts than Behe originally claimed in *Darwin's Black Box*. Ultimately, of course, it is possible to reduce a structure to so few parts that, indeed, removal of any one part will make the structure cease functioning. More important than whether irreducibly complex structures actually occur other than by definition, however, is the critical question of whether they can be produced by natural mechanisms.

Behe answers no, claiming that natural selection, the main mechanism of evolutionary change, is inadequate to the task. He views natural selection as assembling a

Figure 6.1
An "Irreducibly Complex" Mousetrap. Courtesy of Alan Gishlick

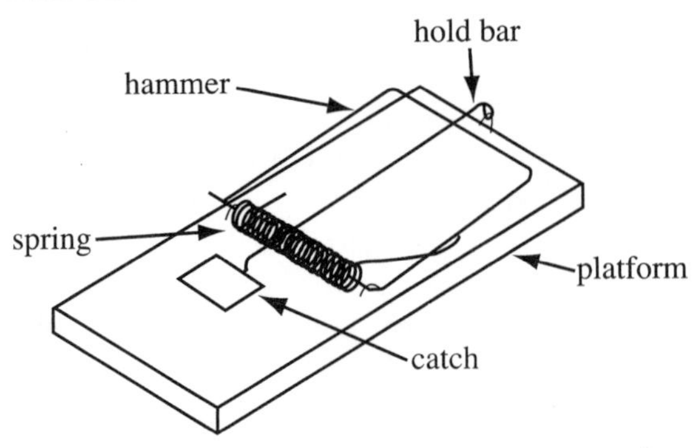

complex structure by stringing together components one at a time, with each addition requiring a selective advantage. Behe's view is that for a structure like the bacterial flagellum, consisting of over 50 proteins and enzymes, it is extraordinarily unlikely that so many elements—by chance—could be assembled one by one, and even more unlikely that there would be selective advantage to each addition. This piece-by-piece assemblage of the flagellum, one enzyme at a time, one after another, is also envisioned by William Dembski (2001), who claims that the probability against this occurring is astronomical; a bacterial flagellum, Dembski and Behe agree, cannot be produced through natural causes.

Critics have noted that Behe presents an incomplete picture of how natural selection operates: it is not the case that components of a complex structure must be added one after another, piece by piece, like stringing beads—the bacterial flagellum need not be a discrete combinatorial object. It is clear from the study of components of a cell that a great deal of borrowing and swapping of bits and pieces takes place: each structure is not composed of unique proteins and enzymes, or even of wholly unique combinations of proteins. The cross-linking proteins of flagella, for example, have other functions elsewhere in the cell. Somewhat fewer than half of the proteins found in the bacterial flagellum are the same as or very similar to those found among other bacteria in a structure called the type-III secretory apparatus, which performs some of the functions of a flagellum. An adaptive advantage for a structural element may exist and cause it to be selected for—but for a different purpose and perhaps in a different cell component than that of the final, supposedly irreducibly complex structure under discussion. Natural selection thus *can* produce complex structures without having to separately string protein after protein together, in which each addition requires a separate action of natural selection. Behe's critics thus have argued that some of the components of an irreducibly complex structure could be assembled separately for some purposes and then combined for other functions—with natural selection being intimately involved during all parts of the process.

There is another way in which natural selection can be more flexible than Behe appears to allow. "Irreducibly complex" structures may indeed have been assembled piece by piece, but the pathway of assembly may not be obvious because of a process evolutionary biologists have called "scaffolding."

Consider an arch made of stone. The keystone, the stone at the top of the arch, must be in place or the arch will fall down; an arch is an irreducibly complex structure. To build it, stonemasons will often build a scaffolding to support the sides of the arch as they build toward the center. When the keystone is laid and the arch is thereafter stable, the scaffolding is removed, leaving the irreducibly complex arch.

An irreducibly complex biochemical or molecular structure may be built in a similar way, in the sense that at an earlier time in the history of the structure, there might have been components that "supported" the function of the structure, much as a scaffolding supports the sides of an arch until the keystone is in place. These "supporting" biochemical components may be made redundant by the addition of more efficient components, much as the arch's scaffolding is made redundant for holding up the sides of the arch once the keystone is laid. The now superfluous components can be removed by natural selection. Without knowing the entire history of the structure, it might seem that all the parts of the structure appeared all at once, fully formed and functional in their final configuration, with no history of earlier, simpler structural predecessors. But just as an arch attains irreducible complexity only at the end of construction, so too do these supposedly irreducibly complex biochemical structures: they actually had a history, though the exact events may not yet have been traced.

But even if natural selection were unable to explain the construction of irreducibly complex structures, does this mean that we must now infer that intelligence is required to produce such structures? Only if there are no other natural causes—known or unknown—that could produce such a structure. Given our current knowledge of the mechanisms of evolution, there is no reason why natural selection could not explain the assembly of an irreducibly complex structure—but it is also the case that a future researcher might come up with an additional mechanism or mechanisms that could explain irreducibly complex structures by some other natural process.

Some scientists have described Behe's approach as an "argument from ignorance" (Blackstone 1997) because the intelligent creator is used as an explanation when a natural explanation is lacking. This is reminiscent of the "God of the gaps" argument, where God's direct action is called upon to explain something that science has not yet explained. "God of the gaps" arguments are rejected by both theologians and scientists. To scientists, using God to explain natural phenomena of any kind violates the practice of methodological naturalism, in which scientific explanations are limited only to natural causes. To theologians, the "God of the gaps" approach creates theological problems of the irrelevance or diminution of God when natural explanations for natural events ultimately replace the direct hand of God. Intelligent Design proponents, however, claim that the issue is not current ignorance of a discoverable natural cause but the *impossibility* of a natural cause.

Behe's idea of irreducible complexity was anticipated in Creation Science; much as in Paley's conception, Creation Science proponents hold that structures "too complex" to have occurred "by chance" require Special Creation. Behe, following ID convention, doesn't mention God directly, but the essence of the irreducible complexity argument is that irreducibly complex structures are evidence for God's direct action. As such, ID verges on being a variety of Progressive Creationism in which God intervenes at intervals to create irreducibly complex structures like DNA, the bacterial flagellum, the blood-clotting cascade, and so on. Although many ID proponents find the Progressive Creationist position attractive, ID is not necessarily wedded to the Progressive Creation position. Some of its proponents suggest that design could have been prearranged (or "front-loaded"). In *Darwin's Black Box*, Behe suggests that perhaps all the irreducibly complex structures of all living things were somehow present in the first living cell, and then appeared through time as various organisms evolved.

> Suppose that nearly four billion years ago the designer made the first cell, already containing all of the irreducibly complex biochemical systems discussed here and many others. (One can postulate that the designs for systems that were to be used later, such as blood clotting, were present but not "turned on." In present-day organisms plenty of genes are turned off for a while, sometimes for generations, to be turned on at a later time.) (Behe 1996: 228)

As noted by the cell biologist Kenneth Miller, such an *über*-cell would somehow have to avoid the mutational drift of genes controlling such structures for billions of years until it was "time" for such structures to appear (Miller 1996: 40). The probability of genes for, say, the bacterial flagellum remaining intact for so long violates much of what we know about the behavior of genes in the absence of natural selection. Such genes tend to accumulate mutations that make the gene nonfunctional.

William Dembski has expanded this concept and proposes that God might have front-loaded everything in the universe in the big bang—all the irreducibly complex structures are merely unfolding like so many homunculi as time passes (Dembski, 2001). In this view, God would not be progressively creating but would have acted only once.

Complex Specified Information: The Design Inference. Dembski's "Design Inference" takes a probability theory approach to distinguish those phenomena in nature that are designed by intelligence from those that are the result of natural cause or chance. Although arguments against evolution based on probability have long been a mainstay in Creation science (Gish 1976; Morris 1974; Perloff 1999), Dembski's Design Inference is at least superficially more impressive, couched as it is in a mathematical idiom. In proposing an "explanatory filter" decision tree, Dembski contends that there are three ways to explain phenomena, based on their frequency of occurrence (see Figure 6.2).

Things that occur commonly or with high predictability can be attributed to the unfolding of natural law. That the moon goes through phases every month can be explained by the passage of the moon around Earth and how our planet's shadow falls upon the moon; it is not necessary to attribute design to this phenomenon. Intermediate probability phenomena may be attributed to chance—even very low probability events will occur some of the time, just by chance alone. But some kinds of low-probability phenomena—Dembski refers to "specified" low-probability events—which are not due to law or to chance compose the class of phenomena that must be attributed to intelligent design. Dembski proposes that *complex specified information* distinguishes intelligently designed phenomena.

Specification is a sort of side information that we add about a phenomenon or event. Consider the explanation for finding an arrow in a bull's-eye. If we see an arrow in a bull's-eye, we might consider that the archer got lucky, but if we see ten arrows in ten bull's-eyes, we attribute this to an archer with a high level of skill. On the other hand, if we knew that the archer shot the arrows first, and then drew the targets around them, we would not attribute the perfect shots to skill. Knowing the targets were present before the arrows is a *specification* or additional information that allows us to attribute the arrows in the bull's-eyes to design rather than chance (or cheating).

Dembski's filter (Figure 6.2) allows the assignment of the causes of some phenomena to natural law, chance, or design, using the combination of probability and specification. High-probability events are stopped by the "natural law" filter; medium- or low, unspecified-probability events are attributed to chance, and only low, specified-probability events are attributed to intelligent design. Dembski's filter is therefore an elimination algorithm: something is explained by design when it is *not* explained by

Figure 6.2
Dembski's "Explanatory Filter" (Dembski 1998: 37).
Courtesy of Alan Gishlick

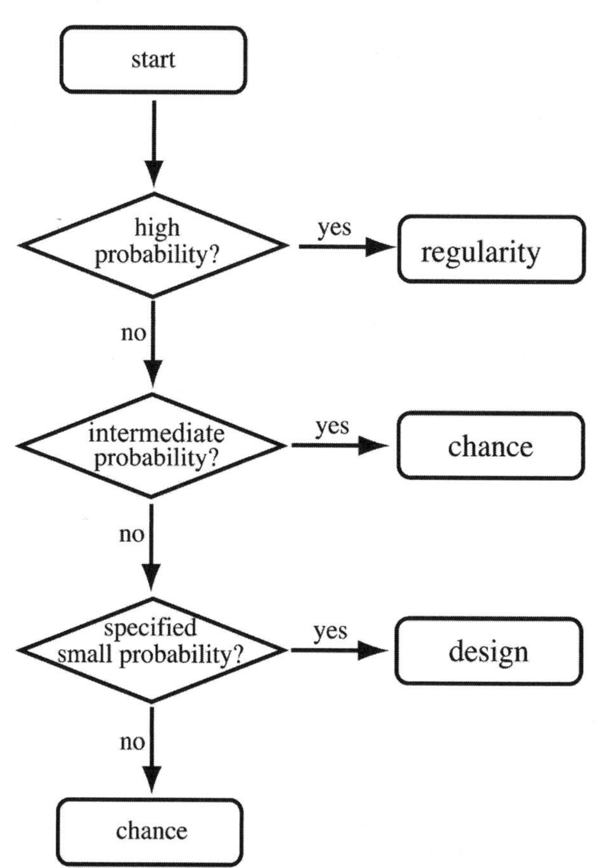

law or chance. But this approach allows false positives where something is attributed to design because of missing or unknown information at the first, natural law level.

For example, let's say that while walking through the forest, we come upon a circle of toadstools that has sprung up overnight. The ring wasn't there yesterday, and in a few days it will largely be in tatters: fungi are fragile things. If this walk were taking place in the 800s in Europe, we peasants would recognize the circle of toadstools as a fairy ring, a place where fairies had had a celebration the night before. Applying Dembski's filter, we would conclude that the sudden and random appearance of the circle, and of course its symmetrical shape, certainly were not the result of natural processes: rings of toadstools crop up with no warning, unlike the phases of the moon. So in the year 800, a fairy ring would pass through the first (regularity) filter. It would also not be attributed to chance, as the likelihood of a fairy ring occurring at a given place is very improbable. However, this low-probability event has a specification, its circular shape. Therefore, following Dembski's filter, we would attribute the appearance of a fairy ring to intelligent design; European peasants of the year 800 knew that fairy rings were the remains of midnight revels held by tiny fairies in the woods.

Perhaps because no one ever found tiny beer cans next to the toadstools, eventually a natural explanation was found for fairy rings: they are the result of one of the ways toadstools reproduce themselves. Some fungi send out underground, threadlike mycelia from a central point, and when circumstances of moisture and temperature are suitable, toadstools form aboveground. These toadstools produce spores that are carried by wind; these spores eventually land and start growing new fungi. Because the mycelia radiate from a center, circles are formed. With additional information, it can be seen that fairy rings actually are not improbable, though many variables are involved in their appearance, and they actually do not appear randomly, but in specific environments. In the twenty-first century, we recognize that fairy rings have a natural explanation; in the ninth century, the circles were explained by design. Because Dembski's filter depends on the extent of scientific knowledge of the time, it thus fails to be a reliable predictor of design by intelligence.

The Problem of Natural Intelligent Cause. As shown in Table 6.1, ID proponents argue that design can be produced both by natural causes (natural selection, for example, has some limited ability to shape organisms to meet some environmental pressures) and by intelligent causes. An intelligently designed phenomenon could be the product of transcendent intelligence such as a creator God, or it could be the product of material agents such as extremely intelligent extraterrestrials—an argument first made, in fact, in the original ID book, *The Mystery of Life's Origin.*

Unfortunately, the dichotomy between "natural" and "intelligent" is artificial, because some of the agents on the "intelligent" side are actually "natural." To a scientist, anything that is the result of matter, energy, and their interactions is a natural phenomenon, whether nonliving phenomena such as stars and rocks or living phenomena such as plants and animals. Material agents such as humans, higher primates, or extraterrestrials (if such beings exist) are therefore natural, as are their behaviors. No one disputes that the behavior of humans and animals can be studied and understood through the application of scientific principles; such behavior is the subject matter of physiology and psychology. We study bird or whale communication,

Table 6.1
Natural and Intelligent Causes: The Intelligent Design View

Natural Cause	Intelligent Cause
Natural Selection	Transcendent Agent (God) Material Agents Humans Higher Primates Extraterrestrials

for example, and attempt to explain it by using general theories and regularities. Behavior is the product of natural entities and is thus itself natural.

To answer whether the *intelligent* behavior of material beings can similarly be studied and explained requires a definition of intelligent behavior—which is not as easy as it seems. Psychologists define intelligence broadly, as having elements of problem-solving and some degree of abstraction. But problem-solving is also a broad category: bees solve the problem of communicating the location of nectar through their waggle dance, which surely is a complex behavior having elements of abstraction (the dance indicates the direction and distance from the hive to the food source), but is this behavior "intelligent"? Intelligent behavior is usually conceived as having some element of choice involved, rather than being the result of largely uncontrolled or genetically wired causes—yet clearly "choice" is a continuum. A bee may be largely hardwired to return to the hive when a source of pollen is discovered and to perform the waggle dance: its genes make it extremely likely that it will respond to a food source by returning directly to the hive. But what if, on the way home, the bee finds a larger source of nectar? Can a "choice" be made not to "report" on the original source but to bring back a message about the second one?

In other instances of behavior, choices clearly are being made. A chimpanzee attains social status through complex interactions with many individuals over a long period of time. Such actions are not genetically hardwired like the waggle dance of the bee, and in fact even involve examples of conscious manipulation of other group members, including efforts to deceive one other (especially over food sources). When primatologists and psychologists study such intelligent behaviors, they attempt to explain them through theoretical principles—in other words, they study them scientifically. Certainly economists, psychologists, and political scientists also study the intelligent (and sometimes unintelligent!) behavior of human beings and attribute all or part of it to patterns. The intelligent behavior of material creatures is therefore natural, and an appropriate subject for scientific investigation.

Therefore, all of the *natural* intelligent agents on the right side of Table 6.1 should be moved to the left side. Furthermore, because the heart of ID is that certain natural phenomena are incapable of explanation through natural cause, it is clear that causation by intelligent material agents should be on the "natural" side of this dichotomy. If material intelligence is moved to the "natural" side of the equation, as in Table 6.2,

Table 6.2
An Alternative View of Natural and Intelligent Causes

Natural Cause	Transcendent Cause
Natural Selection	Transcendent Agent (God)
Material Agents	
Humans	
Higher Primates	
Extraterrestrials	

the ID dichotomy of natural and intelligent must be restructured. What ID proponents wish to label "intelligent" reduces to one agent: God. The dichotomy natural/intelligent is in reality a dichotomy of natural/transcendent. If there is a transcendent agent such as God, who is conceived of as omniscient and omnipotent, then such an agent by definition could not be explained by natural causes and would properly form a dichotomy with natural cause.

Intelligent Design supporters cannot accept this, however, because appealing to transcendent cause is of course a form of religion. They are well aware that the First Amendment does not allow the advocacy of creationism in public schools. Hence it suits ID better to try to combine all forms of intelligent cause into one heterogeneous list—whether or not such a division is empirically or logically defensible.

Intelligent Design supporters are hostile to methodological materialism and propose a new kind of science, "theistic science." This is an alleged subclass of science that deals with the class of scientific problems dealing with "origins" ("origins science"), which are unrepeatable. Such phenomena as the origin of life and the evolution of living things (unspecified) constitute origins science. Although the majority of science may be performed in a methodologically materialistic fashion, explaining only with reference to natural causes, origins science allows (indeed, requires) the occasional invocation of "intelligence"—by which is meant the direct hand of God. Theistic science, then, is a proposal to radically change how we do science by abandoning methodological materialism in favor of allowing a supernatural cause—and still calling the process science. It is not a position which has been embraced by either philosophers of science or scientists (Pennock 1999).

ID's Cultural Renewal Focus

The second focus of ID is "cultural renewal," a term its proponents use to describe the movement's efforts to replace the alleged philosophical materialism of American society with a theistic (especially Christian) religious orientation. Perhaps the most vocal proponent of the cultural renewal focus of ID is retired law professor Phillip Johnson. Although his first antievolution book, *Darwin on Trial*, made only a few references to the purported evils of materialism in American society, subsequent books have been much more evangelical in tone and have strongly and clearly promoted the ID vision for a society with more theistic sensibilities. Conferences (such as "Mere Creation" in 1996) have also promoted sectarian Christian views. Under Johnson's guidance—and taking advantage of his prominence and connections as a professor holding an endowed chair at a leading secular university—the ID movement sought to find acceptance first and foremost from the secular academic community. At about this time, the rapidly expanding ID movement found a new institutional locus beyond the FTE at a conservative think tank in Seattle known as the Discovery Institute. The new movement would have more credibility with academics if it were housed in a more neutral institution than FTE, which has long been associated with evangelical Christianity and thus with Creation Science. The Discovery Institute rapidly replaced the FTE as the hub for ID activities. The FTE Web site, in fact, sounds almost wistful as it describes itself as having "helped to inspire the robust and exciting international movement of Intelligent Design" (www.fteonline.com/about.html, accessed July 16, 2002).

The Discovery Institute (DI) was founded by a politician named Bruce Chapman in 1991 to "promote ideas in the common sense tradition of representative government, the free market and individual liberty" (Discovery Institute mission statement, www.discovery.org/mission.html, accessed August 18, 2002). One program promotes Evangelical Christian political activity (Religion, Liberty and Civic Life), though the majority of the programs echo free market and libertarian themes. The Intelligent Design–promoting Center for Renewal of Science and Culture (CRSC) was announced in a 1996 press release:

> For over a century, Western science has been influenced by the idea that God is either dead or irrelevant. Two foundations recently awarded Discovery Institute nearly a million dollars in grants to examine and confront this materialistic bias in science, law, and the humanities. The grants will be used to establish the Center for the Renewal of Science and Culture at Discovery, which will award research fellowships to scholars, hold conferences, and disseminate research findings among opinion makers and the general public. (Chapman 1996a)

In the Discovery Institute's members' newsletter, Chapman further described the CRSC as having specific religious goals:

> The more you read about the program—and there will be about six books to read from this center in the next four years—the more you will realize the radical assault it makes on the tired and depressing materialist culture and politics of our times, as well as the science behind them. Then, when you start to ponder what society and politics might become under a sounder scientific dispensation, you will become truly inspired. (Chapman 1996b)

The goals of the CRSC have been identified as explicitly religious in other Discovery Institute publications as well: "To defeat scientific materialism and its destructive moral, cultural and political legacies. To replace materialistic explanations with the theistic understanding that nature and human beings are created by God" (CRSC Web site, accessed May 18, 2000). Also, "Accordingly, our Center for the Renewal of Science and Culture seeks to show that science supports the concept of design and meaning in the universe—and that that design points to a knowable moral order" (Chapman 1998: 3).

Until August 2002, the cultural renewal focus was reflected in the name of the Center for Renewal of Science and Culture. In that month, the word "renewal" was dropped from all Web pages, and the CRSC became the Center for Science and Culture (CSC). One may speculate that "cultural renewal" may have been too reminiscent of the goals of twentieth-century Creation Science, distracting attention from the scholarly focus: scientific and other scholarly organizations do not typically have as their goal the "renewal" of culture (Holden 2002).

Although ID proclaims itself a scholarly movement, its cultural renewal focus is fundamentally incompatible with the openness and flexibility demanded by a scientific theoretical perspective. Enamored of an ideological, political, or social goal, it is all too easy to misrepresent or ignore empirical data when they do not support the goal; certainly Creation Science is infamous for doing so (Scott 1993). A few ID proponents appear to be aware that the scholarly aspect of ID has taken a backseat to the political and the ideological. Bruce Gordon has been especially eloquent on this issue, writing that:

> . . . design-theoretic research has been hijacked as part of a larger cultural and political movement. In particular, the theory has been prematurely drawn into discussions of public science education, where it has no business making an appearance without broad recognition from the scientific community that it is making a worthwhile contribution to our understanding of the natural world. (Gordon 2001: 9)

Gordon also commented, "If design theory is to make a contribution in science, it must be worth pursing on the basis of its own merits, not as an exercise in Christian 'cultural renewal,' the weight of which it cannot bear" (Gordon 2001: 9).

Intelligent Design's Content Problem

A criticism of the ID movement brought by both scientists and Young Earth Creationists is the lack of empirical content in ID "theory." As detailed by the Young Earth Creationist Carl Wieland, on the *Answers in Genesis* Web site:

> They generally refuse to be drawn on the sequence of events, or the exact history of life on Earth or its duration, apart from saying, in effect, that it "doesn't matter." However, this is seen by the average evolutionist as either absurd or disingenuously evasive—the arena in which they are seeking to be regarded as full players is one which directly involves historical issues. In other words, if the origins debate is not about a "story of the past," what is it about? (Wieland 2002)

Most ID proponents accept an ancient age of the universe and Earth, but there are some prominent ID supporters, such as Paul Nelson and John Mark Reynolds, who joined the ID camp from the Creation Science side of antievolutionism. These Creation Science adherents reject evolution altogether, whereas some ID supporters such as Michael Behe have gone so far as to accept common ancestry of humans and apes

(Miller 1996). The range of scientific opinion within the ID camp, therefore, runs from Young Earth Creationism to mild forms of theistic evolution, although Dembski and others have declared theistic evolution to be incompatible with ID (Dembski 1995). The ID movement surely is a proverbial "big tent," though it remains to be seen whether the differences among the tent's occupants will be reconcilable if ID takes any specific empirical positions on what Wieland has called the "story of the past" (Scott 2001).

However, common to all ID proponents is the rejection of "Darwinism." In ID literature, "Darwinism" becomes an epithet, though it is not always clear in any given passage exactly what is meant by "Darwinism." In evolutionary biology, "Darwinism" usually refers to the general idea of evolution by natural selection; it may specifically refer to the ideas held by Darwin in the nineteenth century. Usually the term is not used for modern evolutionary theory, which, because it goes well beyond Darwin to include subsequent discoveries and understandings, is more frequently referred to as "neo-Darwinism," or just "evolutionary theory." Evolutionary biologists hardly ever use "Darwinism" as a synonym for evolution, though it occasionally is used this way by historians and philosophers of science. In ID literature, however, "Darwinism" can mean Darwin's ideas, natural selection, neo-Darwinism, post-neo-Darwinian evolutionary theory, evolution itself, or materialist ideology inspired by "Godless evolution."

The public, on the other hand, is unlikely to make these distinctions, instead equating "Darwinism" with evolution (common descent). For decades, Creation Science proponents have cited the controversies among scientists over *how* evolution occurred—including the specific role of natural selection—in their attempts to persuade the public that evolution itself—the thesis of common ancestry—was not accepted by scientists, or at least was in dispute. Within the scientific community, of course, there are lively controversies, including over how much of evolution is explained by natural selection and how much by additional mechanisms such as those being discovered in evolutionary developmental biology ("evo-devo"). No one says natural selection is unimportant; no one says that additional mechanisms are categorically ruled out. But these technical arguments go well beyond the understanding of laypeople and are easily used to promote confusion over *whether* evolution occurred.[1] Intelligent Design proponents similarly exploit public confusion about "Darwinism" to promote doubt about evolution.

The rejection of Darwinism, however, is not only the glue that holds the disparate ID proponents together; it is also central to their movement. The natural selection mechanism of evolutionary change has long vexed conservative Christians who have difficulty reconciling it with the concept of a loving, all-good Creator who is personally involved with creation. Concern with theodicy (the problem of evil) in Christian theology of course predated Darwin's discoveries, but there is no escaping that natural selection has implications for certain Christian views. There are many ways that Christian theologians have integrated the natural selection mechanism into different views of God (Hewlett and Peters 2003), though these compromises, along with theistic evolution, are rejected by the ID proponents. Natural selection is acceptable—in fact, undeniable—on the level of a population of organisms: neither ID nor Creation Science proponents deny the ability of natural selection to lengthen bird beaks or produce antibiotic-resistant bacteria or pesticide-resistant insects. But for God to use the wasteful and cruel mechanism of natural selection to produce the diversity of living things today is theologically unacceptable to many in the ID camp. Natural selection, they believe, works only at the "local" level and lacks the creative power to produce new body plans and bring about significant evolutionary changes among groups.

Is Intelligent Design "Creationist"?

Intelligent Design proponents do not refer to themselves as creationists, associating that term, as many do, with the followers of Henry M. Morris. Indeed, most ID proponents do not embrace the Young Earth, Flood Geology, and sudden creation tenets associated with YEC. Yet by Phillip Johnson's definition, ID proponents arguably are creationists:

> "Creationism" means belief in creation in a more general sense. Persons who believe that the earth is billions of years old, and that simple forms of life evolved gradually to become more complex forms including humans, are "creationists" if they believe that a supernatural Creator not only initiated this process but in some meaningful sense *controls* it in furtherance of a purpose. (Johnson 1991: 4)

Phillip Johnson contends that the scientific data and theory supporting evolution are weak, and that evolution persists as a scientific idea only because it reinforces philosophical materialism. To him and most other ID proponents, the most important issue in the creation/evolution controversy is whether the universe came to its present state "through purposeless, natural processes known to science" (Johnson 1990: 30) or whether God had meaningful involvement with the process. Intelligent Design proponents clearly believe that God is an active participant in creation, though they are divided as to whether this activity takes the form of front-loading all outcomes at the big bang, episodic intervention of the progressive creationism form, or other, less well-articulated possibilities. Theistic evolution, however, is ruled out.

"EVIDENCE AGAINST EVOLUTION"

The *Edwards* decision, as mentioned, rejected equal time for creationism and evolution but allowed secular, scientific alternatives to evolution to legally be taught. Antievolutionists generated Abrupt Appearance Theory and Intelligent Design Theory because scientific alternatives to evolution were not found in the scientific community. But one of the consistent messages of Creation Science is that evolution is a weak scientific theory with much evidence against it. Most of Creation Science consists of alleged arguments against evolution, with few positive empirical claims of its own other than Flood Geology; this observation was also made in *McLean v. Arkansas*.

Justice Scalia's dissent to the *Edwards* decision suggested a strategy of "balancing" evolution by teaching the "evidence against evolution":

> The people of Louisiana, including those who are Christian fundamentalists, are quite entitled, as a secular matter, to have whatever scientific evidence there may be against

> evolution presented in their schools, just as Mr. Scopes was entitled to present whatever scientific evidence there was for it. (Scalia dissent, conclusion)

In the spring of 1996, a bill was narrowly defeated in Ohio to mandate the teaching of evidence against evolution. The proposed law read, in part:

> Whenever a theory of the origin of humans, other living things, or the universe that might commonly be referred to as "evolution" is included in the instructional program provided by any school district or educational service center, both evidence and arguments supporting or consistent with the theory and evidence and arguments problematic for, inconsistent with, or not supporting the theory shall be included.

Also during the spring of 1996, Georgia's legislature considered similar legislation, which stated, in part: "Teachers shall have the right to present and critique any and all scientific theories about such origins and all facets thereof." In these and other bills, it is assumed that "evidence against evolution" exists. In some bills, such as a 2001 Montana bill (HB 588), "evidence against evolution" is asserted to exist:

> WHEREAS, compelling evidence exists in support of divergent scientific conclusions and validly competing theories of origin and the spirit of science requires that students be impartially exposed to all evidence, including scientific information tending to prove and disprove each theory . . .

A 2001 Arkansas bill (HB 2548) forbade educational agencies to use public funds to purchase textbooks or other instructional materials lacking antievolution arguments from Intelligent Design and traditional Creation Science sources:

> No state agency, city, county, school district or political subdivision shall use any public funds to provide instruction or purchase books, documents or other written material which it knows or should have known contain descriptions, conclusions, or pictures designed to promote the false evidences set forth in subsection (d) of this section.

Section (d) of the bill listed supposed examples of evolutionary fraud taken from ID proponent Jonathan Wells's book, *Icons of Evolution*, Haeckel's embryos, the Miller–Urey experiment, *Archaeopteryx* (the ancient bird), and the peppered moth. From traditional Creation Science (Chick 2000) the bill's author took such staples as Piltdown man, Nebraska man, and Neanderthal man—all claimed to be fraudulent, though only the first actually was. Elsewhere in the bill were references to other Creation Science claims such as gaps in the fossil record, falsity of the geological column, and flaws in radiometric dating. Even if one were not familiar with Creation Science literature, the reference to "evidences"—a term from Christian apologetics rather than science—would reveal the inspiration for this bill.

In addition to assuming that scientific evidence against evolution exists, such bills—like the equal time bills they supplanted—appeal to the American public's appreciation of fairness. The American political tradition of local decision-making (e.g., by town councils or local school boards) encourages a wide variety of voices to contend for influence and authority. Part of the political as well as cultural tradition is for all voices to have an opportunity to be heard, even if later rejected. This is enshrined in the First Amendment's Free Speech and Assembly clauses and manifest even in journalistic traditions where the reporter is expected to present "both views" of a controversy. As will be discussed elsewhere (chapter 10), the "fairness" approach, though culturally very powerful, is misapplied in the realm of science, which actually is highly discriminatory! In science, a variety of views may indeed be heard, but once rejected, there must be a compelling reason for them to be taken seriously again. Scientific claims for the world and its inhabitants coming into being suddenly, at one time, in their present form, have not been taken seriously since the end of the eighteenth century, and it is unfair to pretend to students that this view is a viable scientific option in the twenty-first century.

"Evidence against evolution" is emerging as a popular antievolution approach. It is attractive to legal specialists among the antievolutionists because it appears to avoid the Establishment Clause of the First Amendment by not obviously promoting religion. Presenting evidence against evolution per se is only bad science, which the First Amendment does not forbid.

"JUST A THEORY" DISCLAIMERS

Another Neocreationist approach is to denigrate evolution by requiring that it be distinguished from all other scientific explanations as a "theory." Often such efforts are coupled with requirements that disclaimers, often claiming that evolution is "just a theory," be included in textbooks or read to students. As discussed in chapter 1, theory is used in science to mean "explanation," whereas in casual use, it is often a synonym of "guess" or "hunch." Scientific theories are far from guesses: there are many explanations in science, and the best ones are elevated to theories. When school boards or state legislatures attempt to single out evolution as "just a theory," it is clear that they are not using this term in its scientific sense. But such disclaimers and policies have the net effect of drawing attention to evolution as a particularly controversial subject, making it less likely that evolution will be taught.

Efforts to require disclaimers for evolution began in Texas, when in 1974 the state board of education required that all biology textbooks bought in the state treat evolution "as a theory" and not "factually verifiable." "Furthermore, each textbook must carry a statement on an introductory page that any material on evolution included in the book is clearly presented as theory rather than verified" (Mattox et al. 1984: 1). Although in 1984 the Texas attorney general opined that the Texas disclaimer was illegal (see chapter 9), other states and communities have regularly proposed and passed such evolution-only disclaimers.

Perhaps the best-known and most far-reaching has been the Alabama textbook disclaimer of 1995. The state board of education decided that all biology textbooks must have a disclaimer on evolution pasted into them, which began by identifying evolution as a theory rather than a fact.

> This textbook discusses evolution, a controversial theory some scientists present as a scientific explanation for the origin of living things, such as plants, animals and humans. No one was present when life first appeared on earth, therefore, any statement about life's origins should be considered as theory, not fact.

The Alabama disclaimer spread to some school districts in neighboring Georgia, although it failed at the state legislative level there. It also appeared in 1998 in proposed legislation in Washington state. Although unsuccessful, the persistent Washington legislator continued to introduce this legislation in 2001 and 2002, equally unsuccessfully. Alabama disclaimer legislation was also introduced in Oklahoma and in Mississippi in 2003 but did not pass in either state. Ironically, Alabama rewrote its disclaimer in 2001, reducing the amount of scientific error it contained (Meikle 2001), but the "old" disclaimer is still being proposed in other states and communities.

The vast majority of "theory, not fact" policies and disclaimers do not pass, but the publicity given to them contributes to the general perception that evolution is somehow less valid than other scientific subjects. A disclaimer that was passed by the Tangipahoa, Louisiana, board of education in 1994 singled out evolution for special treatment. Teachers were directed to read the disclaimer to students before discussing evolution or assigning readings. The disclaimer read in part:

> It is hereby recognized by the Tangipahoa Parish Board of Education, that the lesson to be presented, regarding the origin of life and matter, is known as the Scientific Theory of Evolution and should be presented to inform students of the scientific concept and not intended to influence or dissuade the Biblical version of Creation or any other concept.
>
> It is further recognized by the Board of Education that it is the basic right and privilege of each student to form his/her own opinion or maintain beliefs taught by parents on this very important matter of the origin of life and matter. Students are urged to exercise critical thinking and gather all information possible and closely examine each alternative toward forming an opinion.

The Tangipahoa disclaimer was challenged in Federal District Court, which ruled in 1997 that the purpose of the regulation was to promote religion and that the second paragraph's attempt to present the disclaimer as having the purpose of promoting critical thinking was a sham. This determination was made based on the facts of the case, in which it was apparent from minutes and other reports of the board of education that the legislation was intended to promote specific sectarian (biblical Christian) views. Significantly, however, the court noted that it was possible that some form of disclaimer could be constitutional—although the Tangipahoa disclaimer, with its specific mention of the Bible, was not.

It is anticipated that new disclaimers will continue to be proposed, as other, if more subtle, forms of antievolutionism.

NOTE

1. Note that none of the active proponents of either ID or Creation Science are contributing to the scientific discussion of these points: instead of debating evolution theory at professional scientific conferences or in journals, they do no research on evolutionary biology but merely report on the work of other scientists (e.g., Johnson 1991; Wells 2000), often distorting it severely in the process (Branch 2002; Coyne 1996, 2002; Padian and Gishlick 2002; Scott 2001; Scott and Sager 1992; Gould 1992). They prepare their articles and books for the general reader rather than the scientific public.

REFERENCES CITED

Behe, Michael. 1996. *Darwin's Black Box: The Biochemical Challenge to Evolution*. New York: The Free Press.

Bird, Wendell. 1987. *The Origin of Species Revisited: The Theories of Evolution and Abrupt Appearance*, vol. 1. New York: Philosophical Library.

Blackstone, Neil W. 1997. Argumentum ad Ignorantiam: A Review of "Darwin's Black Box: The Biochemical Challenge to Evolution." *Quarterly Review of Biology* 72 (4): 445–447.

Branch, Glenn. 2002. Analysis of the Discovery Institute Bibliography. *Reports of the National Center for Science Education*, 22 (4): 12–18, 23–24 .

Chapman, Bruce. 1996a. Press Release. Seattle, WA: Discovery Institute, October 10.

Chapman, Bruce. 1996b. Ideas Whose Time Is Come. *Discovery Institute Journal*, August.

Chapman, Bruce. 1998. Letter from the President. Seattle, WA: Discovery Institute.

Chick, Jack T. 2000. *Big Daddy*. Pamphlet from Chick Publications, Ontario, CA.

Coyne, Jerry A. 1996. God in the Details. *Nature* 383: 227–228.

Coyne, Jerry A. 2002. Creationism by Stealth. *Nature* 410: 745–746.

Dembski, William. 1995. What Every Theologian Should Know About Creation, Evolution, and Design. *Center for Interdisciplinary Studies Transactions* 3 (2): 1–8.

Dembski, William. 2001. *No Free Lunch: Why Specified Complexity Cannot Be Purchased Without Intelligence*. Lanham, MD: Rowman & Littlefield.

Gish, Duane T. 1976. *Origin of Life: Critique of Early Stage Chemical Evolution Theories*. San Diego: Institute for Creation Research.

Gordon, Bruce. 2001. Intelligent Design Movement Struggles with Identity Crisis. *Research News and Opportunities in Science and Theology*, January: 9.

Gould, Stephen Jay. 1992. Impeaching a Self-Appointed Judge. *Scientific American*, July: 118–121.

Hewlett, Martinez, and Ted Peters. 2003. *Evolution from Creation to New Creation*. Nashville, TN: Abingdon Press.

Holden, Constance. 2002. Design's Evolving Image. *Science* 297: 1991.

Hoyle, Fred, and Chandra Wickramasinghe. 1979. *Diseases from Space*. New York: Harper & Row.

Johnson, Phillip E. 1990. *Evolution as Dogma: The Establishment of Naturalism*. Dallas, TX: Haughton.

Johnson, Phillip E. 1991. *Darwin on Trial*. Washington, DC: Regnery Gateway.

Kennedy, J. 1992. Teaching of Creationism Splits Louisville Parents. *The Repository*, September 30, B4.

Mattox, Jim, Tom Green, David R. Richards, and Rick Gilpin. 1984. Opinion no. JM-134. Austin: Office of the Attorney General, State of Texas. Available from http://www.oag.state.tx.us/opinions/op47mattox/jm-0134.htm.

Meikle, Eric. 2001. State Board of Education Adopts Another Evolution Disclaimer. Online publication at http://www.ncseweb.org/resources/news/2001/AL/123_state_board_of_education_adopt_11_8_2001.asp, accessed October 4, 2003.

Miller, Kenneth R. 1996. Review of: "Darwin's Black Box: The Biochemical Challenge to Evolution." *Creation/Evolution* 16 (2): 36–40.

Morris, Henry M., ed. 1974. *Scientific Creationism*. San Diego: Creation-Life Publishers.

Padian, Kevin, and Alan D. Gishlick. 2002. The Talented Mr. Wells. *Quarterly Review of Biology* 77 (1): 33–37.

Pennock, Robert T. 1999. *Tower of Babel: The Evidence Against the New Creationism*. Cambridge, MA: MIT Press.

Perloff, James. 1999. *Tornado in a Junkyard: The Relentless Myth of Darwinism*. Arlington, MA: Refuge Books.

Scott, Eugenie C. 1989. New Creationist Book on the Way. *Reports of the National Center for Science Education* 9 (2): 21.

Scott, Eugenie C. 1993. The Social Context of Pseudoscience. In *The Natural History of Paradigms*, edited by J. H. Langdon and M. E. McGann. Indianapolis: University of Indianapolis Press.

Scott, Eugenie C. 2001. The Big Tent and the Camel's Nose. *Reports of the National Center for Science Education* 21 (2): 39. On-line at http://www.ncseweb.org/resources/articles/4397_the_big_tent_and_the_camel39_2_2_2001.asp.

Scott, Eugenie C., and Thomas C. Sager. 1992. "Darwin on Trial": A Review. *Creation/Evolution* 12 (2): 47–56.

Thaxton, Charles B., Walter L. Bradley, and Roger L. Olsen. 1984. *The Mystery of Life's Origin: Reassessing Current Theories*. New York: Philosophical Library.

Weiland, Carl. 2002. *AIG's Views on the Intelligent Design Movement*. Answers in Genesis. Accessed September 4, 2002. Available from http://www.answersingenesis.org/docs2002/0830_IDM.asp.

Wells, Jonathan. 2000. *Icons of Evolution: Science or Myth?* Washington, DC: Regnery.

LEGAL CASES CITED

Edwards v. Aguillard 482 U.S. 578 (1987).

McLean v. Arkansas Board of Education 529 F. Supp. 1255 (E.D. La. 1982).

Part III

Selections from the Literature

The social movement known as creationism has had a long history and a variety of manifestations through time. To understand it takes considerable study. References for further reading will be presented in the final section of this book, but suffice it to say that there is an extensive literature on the topic. Additionally, there are many active creationists/antievolutionists today who maintain Web sites, sponsor lecture tours, and communicate with the public in other ways.

In part III, I present selections from the antievolutionist literature and responses from the pro-evolution side. I do not present evidence for evolution; as emphasized throughout this book, evolution is the consensus view of the scientific community, supported by overwhelming evidence from a variety of scientific fields. The best way to begin to learn about the evidence is to take courses at the university level (they are rarely offered in high school) or to study popular—or better, scientific—sources presenting the evidence and theory of this science, some of which are included as references in other chapters and in part IV. As discussed in chapter 2, evolution is included among a variety of sciences, including astronomy, geology, biochemistry, biology, and anthropology; because this book concentrates on the creationism/evolution controversy, I have made antievolutionism the focus of selections from the literature.

Part III is organized topically, beginning with the physical sciences in chapter 7 and moving to biology in chapter 8 and legal issues in chapter 9. Educational issues are taken up in chapter 10, followed by religious issues in chapter 11 and topics relevant to the philosophy of science in chapter 12.

CHAPTER 7

Cosmology, Astronomy, Geology

INTRODUCTION

Creationists promote their position that creationism is scientific through the sciences of astronomy, geology, and cosmology, the study of the origins and development of the universe. Chemistry and physics tend to be cited primarily through their relationship to cosmology: standard cosmological theory, for example, claims that atomic elements evolved in nuclear reactions in stars; Special Creationism contends that they were created by God all at one time. The most frequently encountered creationist claim from the physical sciences is that the second law of thermodynamics makes evolution impossible.

Much of the Young Earth Creationist (YEC) literature relevant to the physical sciences concerns arguments for a young Earth. The vast majority of creationist arguments are attacks on evolution rather than positive evidence supporting the inference of a young Earth. Carefully selected articles from the scientific literature are claimed to be incompatible with evolution, and, with evolution thus disproved, Special Creationism remains as the logical alternative. For example, calculations of the influx of meteoritic dust on the moon and the current depth of this dust imply to YECs that the moon (and therefore Earth) must be young, or else the moon's surface would be covered with hundreds of feet of dust. Many of these arguments are technical, and although those that have been examined by the scientific community have been countered, the refutations require more background information than most readers of this small book likely will have. I have therefore chosen instead to present one of the positive creationist arguments, the "vapor canopy theory"; both the argument and its refutation require only some basic scientific information.

Also supporting the contention of a young Earth is the creationist argument that radiometric dating is scientifically flawed, and therefore standard scientific claims for a 4-billion-year-old Earth are unsupported. However, many of the classic arguments are being abandoned by some creationists, such as "Darwin's deathbed confession," "NASA's discovery of a 'missing day,'" "women having one more rib than men," and others. In the selections countering creationist claims, I present criticisms by creationists of some of these old chestnuts. Although many individual creationists hold these positions, they clearly are not promoted by one of the major creationist organizations, Answers in Genesis, from which these criticisms come.

Intelligent Design (ID), on the other hand, attempts to avoid the question of the age of Earth and, indeed, most other fact claims about the nature of the universe. The exception is the "anthropic principle," a cosmological concept. At heart, the anthropic argument is a design argument, proposing that God created a perfectly tuned universe in order that humankind would evolve. As such, it is embraced by both Young Earth and Intelligent Design creationists.

•••••••••••••••••••••••••••••

SELECTION SUPPORTING CREATIONIST VIEWS: GENERAL

The Tenets of Creationism

Creationism can be studied and taught in any of three basic forms, as follows:

(1) Scientific creationism (no reliance on Biblical revelation, utilizing only scientific data to support and expound the creation model).
(2) Biblical creationism (no reliance on scientific data, using only the Bible to expound and defend the creation model).
(3) Scientific Biblical creationism (full reliance on Biblical revelation but also using scientific data to support and develop the creation model).

These are not contradictory systems, of course, but supplementary, each appropriate for certain applications. For example, creationists should not advocate that Biblical creationism be taught in public schools, both because of judicial restrictions against religion in such schools and also (more importantly) because teachers who do not believe the Bible should not be asked to teach the Bible. It is both legal and desirable, however, that scientific creationism be taught in public schools as a valid alternative to evolutionism.

In a Sunday School class, on the other hand, dedicated to teaching the Scriptures and "all the counsel of God," Biblical creationism should be strongly expounded and emphasized as the foundation of all other doctrine. In a Christian school or college, where the world of God is studied in light of the Word of God, it is appropriate and very important to demonstrate that Biblical creationism and scientific creationism are fully compatible, two sides of the same coin, as it were. The creation revelation in Scripture is thus supported by all true facts of nature; the combined study can properly be called scientific Biblical creationism. All three systems, of course, contrast sharply and explicitly with the evolution model.

The evolution and creation models, in their simplest forms, can be outlined as follows:[1]

Evolution Model	Creation Model
1. Continuing naturalistic origin	1. Completed supernaturalistic origin
2. Net present increase in complexity	2. Net present decrease in complexity
3. Earth history dominated by uniformitarianism	3. Earth history dominated by catastrophism

The evolution model, as outlined above, is in very general terms. It can be expanded and modified in a number of ways to correspond to particular types of evolutionism (atheistic evolution, theistic evolution, Lamarckianism, neo-Darwinism, punctuated equilibrium, etc.).

The same is true of the creation model, with the Biblical record giving additional specific information which could never be determined from science alone. The three key items in the creation model above are then modified as follows:

Biblical Creation Model

1. Creation completed by supernatural processes in six days.
2. Creation in the bondage of decay because of sin and the curse.
3. Earth history dominated by the great flood of Noah's day.

Creationists, however, do not propose that the public schools teach six-day creation, the fall of man, and the Noachian flood. They do maintain, however, that they should teach the evidence for a complex completed creation, the universal principle of decay (in contrast to the evolutionary assumption of increasing organization), and the worldwide evidences of recent catastrophism. All of these are implicit in observable scientific data, and should certainly be included in public education.

Both the scientific creation model and the Biblical creation model can be considerably expanded to incorporate many key events of creation and earth history, in terms of both scientific observation on the one hand and Biblical doctrine on the other. These can, in fact, be developed as a series of formal tenets[2] of scientific creationism and Biblical creationism, respectively, as listed below:

Tenets of Scientific Creationism

1. The physical universe of space, time, matter, and energy has not always existed, but was supernaturally created by a transcendent personal Creator who alone has existed from eternity.

2. The phenomenon of biological life did not develop by natural processes from inanimate systems but was specially and supernaturally created by the Creator.

3. Each of the major kinds of plants and animals was created functionally complete from the beginning and did not evolve from some other kind of organism. Changes in basic kinds since their first creation are limited to "horizontal" changes (variations) within the kinds, or "downward" changes (e.g., harmful mutations, extinctions).

4. The first human beings did not evolve from an animal ancestry, but were specially created in fully human form from the start. Furthermore, the "spiritual" nature of man (self-image, moral consciousness, abstract reasoning, language, will, religious nature, etc.) is itself a supernaturally created entity distinct from mere biological life.

5. Earth pre-history, as preserved especially in the crustal rocks and fossil deposits, is primarily a record of catastrophic intensities of natural processes, operating largely within uniform natural laws, rather than one of uniformitarian process rates. There is therefore no a priori reason for not considering the many scientific evidences for a relatively recent creation of the earth and the universe, in addition to the scientific evidences that most of the earth's fossiliferous sediments were formed in an even more recent global hydraulic cataclysm.

6. Processes today operate primarily within fixed natural laws and relatively uniform process rates. Since these were themselves originally created and are daily maintained by their Creator, however, there is always the possibility of miraculous intervention in these laws or processes by their Creator. Evidences for such intervention must be scrutinized critically, however, because there must be clear and adequate reason for any such action on the part of the Creator.

7. The universe and life have somehow been impaired since the completion of creation, so that imperfections in structure, disease, aging, extinctions and other such phenomena are the result of "negative" changes in properties and processes occurring in an originally perfect created order.

8. Since the universe and its primary components were created perfect for their purposes in the beginning by a competent and volitional Creator, and since the Creator does remain active in this now-decaying creation, there does exist ultimate purpose and meaning in the universe. Teleological considerations, therefore, are appropriate in scientific studies whenever they are consistent with the actual data of observation, and it is reasonable to assume that the creation presently awaits the consummation of the Creator's purpose.

9. Although people are finite and scientific data concerning origins are always circumstantial and incomplete, the human mind (if open to the possibility of creation) is able to explore the manifestation of that Creator rationally and scientifically, and to reach an intelligent decision regarding one's place in the Creator's plan. . . .

NOTES

1. See *Scientific Creationism*, ed. Henry M. Morris (San Diego: C.L.P. Publishers, 1974), p. 12.

2. These tenets have recently been adopted by the staff of the Institute for Creation Research and incorporated permanently in its by laws.

Selection excerpted from:

Morris, Henry M. 1980. The Tenets of Creationism. *Impact*, July: 1–4.

● ●

SELECTION COUNTERING CREATIONIST CLAIMS: GENERAL

Moving Forward: Arguments We Think Creationists Shouldn't Use

"Darwin Recanted on His Deathbed"

Many people use this story, originally from a Lady Hope. However, it is almost certainly not true, and there is no corroboration from those who were closest to him, even from Darwin's wife, Emma, who never liked evolutionary theory. Also, even if true, so what? If Ken Ham recanted Creation, would that disprove it? So there is no value to this argument whatever (Grigg 1995).

"Moon Dust Thickness Proves a Young Moon"

For a long time, creationists claimed that the dust layer on the moon was too thin if dust had truly been falling on it for billions of years. They based this claim on early estimates—by evolutionists—of the influx of moon dust, and worries that the moon landers would sink into this dust layer. But these early estimates were wrong, and by the time of the Apollo landings, most in NASA were not worried about sinking. So the dust layer thickness can't be used as proof of a young moon (or of an old one either).

"Women Have One More Rib Than Men"

AiG [Answers in Genesis] has long pointed out the fallacy of this statement. Dishonest skeptics wanting to caricature creation also use it, in reverse. The removal of a rib would not affect the genetic instructions passed on to the offspring, any more than a man who loses a finger will have sons with nine fingers.

Note also that Adam wouldn't have had a permanent defect, because the rib is the one bone that can regrow if the surrounding membrane (periosteum) is left intact (Wieland 1999).

"NASA Computers, in Calculating the Positions of Planets, Found a Missing Day and 40 Minutes, Proving Joshua's 'Long Day' and Hezekiah's Sundial Movement of Joshua 10 and 2 Kings 20"

This is a hoax. Essentially the same story, now widely circulated on the Internet, appeared in the somewhat unreliable 1936 book *The Harmony of Science and Scripture* by Harry Rimmer. Evidently an unknown person embellished it with modern organization names and modern calculating devices.

Also, the whole story is mathematically impossible—it requires a fixed reference point before Joshua's long day. In fact we would need to cross-check between both astronomical and historical records to detect any missing day. And to detect a missing 40 minutes requires that these reference points be known to within an accuracy of a few minutes. It is certainly true that the timing of solar eclipses observable from a certain location can be known precisely. But the ancient records did not record time that precisely, so the required cross-check is simply not possible. Anyway, the earliest historically recorded eclipse occurred in 1217 B.C., nearly two centuries after Joshua. So there is no way the missing day could be detected by any computer.

Note that discrediting this myth doesn't mean that the events of Joshua 10 didn't happen. Features in the account support its reliability, e.g., the moon was also slowed down. This was not necessary to prolong the day, but this would be observed from Earth's reference frame if God had accomplished this miracle by slowing Earth's rotation (Grigg 1997).

REFERENCES CITED

Grigg, R. 1995. Did Darwin Recant? *Creation* 18 (1): 36–37.

Grigg, R. 1997. Joshua's Long Day: Did It Really Happen—and How? *Creation* 19 (3): 35–37.

Wieland, C. 1999. Regenerating Ribs: Adam and That "Missing Rib." *Creation* 21 (4): 46–47.

Selection excerpted from:

Sarfati, Jonathan. 2002. Moving Forward: Arguments We Think Creationists Shouldn't Use. *Creation* 24 (2): 20–24.

●●●●●●●●●●●●●●●●●●●●●●●●●●●●

THE SECOND LAW OF THERMODYNAMICS PRECLUDES EVOLUTION: CREATIONIST VIEWS

The Scientific Case Against Evolution: A Summary. Part II

. . . The main scientific reason why there is no evidence for evolution in either the present or the past (except in the creative imagination of evolutionary scientists) is because one of the most fundamental laws of nature precludes it. The law of increasing entropy—also known as the second law of thermodynamics—stipulates that all systems in the real world tend to go "downhill," as it were, toward disorganization and decreased complexity.

This law of entropy is, by any measure, one of the most universal, best-proved laws of nature. It applies not only in physical and chemical systems, but also in biological and geological systems—in fact all systems, without exception.

> No exception to the second law of thermodynamics has ever been found—not even a tiny one. Like conservation of energy (the "first law"), the existence of a law so precise and so independent of details of models must have a logical foundation that is independent of the fact that matter is composed of interacting particles. (Lieb and Yngvason 2000)

The author of this quote is referring primarily to physics, but he does point out that the second law is "independent of details of models." Besides, practically all evolutionary biologists are reductionists—that is, they insist that there are no "vitalist" forces in living systems, and that all biological processes are explicable in terms of physics and chemistry. That being the case, biological processes also must operate in accordance with the laws of thermodynamics, and practically all biologists acknowledge this.

Evolutionists commonly insist, however, that evolution is a fact anyhow, and that the conflict is resolved by noting that the earth is an "open system," with the incoming energy from the sun able to sustain evolution throughout the geological ages in spite of the natural tendency of all systems to deteriorate toward disorganization. That is how an evolutionary entomologist has dismissed W. A. Dembski's impressive recent book, *Intelligent Design*. This scientist defends what he thinks is "natural processes' ability to increase complexity" by noting what he calls a "flaw" in "the arguments against evolution based on the second law of thermodynamics." And what is this flaw?

> Although the overall amount of disorder in a closed system cannot decrease, local order within a larger system can increase even without the actions of an intelligent agent. (Johnson 2000: 274)

This naive response to the entropy law is typical of evolutionary dissimulation. While it is true that local order can increase in an open system if certain conditions are met, the fact is that evolution does not meet those conditions. Simply saying that the earth is open to the energy from the sun says nothing about how that raw solar heat is converted into increased complexity in any system, open or closed.

The fact is that the best known and most fundamental equation of thermodynamics says that the influx of heat into an open system will increase the entropy of that system, not decrease it. All known cases of decreased entropy (or increased organization) in open systems involve a guiding program of some sort and one or more energy conversion mechanisms.

Evolution has neither of these. Mutations are not "organizing" mechanisms, but disorganizing (in accord with the second law). They are commonly harmful, sometimes neutral, never beneficial (at least as far as observed mutations are concerned). Natural selection cannot generate order, but can only "sieve out" the disorganizing mutations presented to it, thereby conserving the existing order, but never generating new order. In principle, it may be barely conceivable that evolution could occur in open systems, in spite of the tendency of all systems to disintegrate sooner or later. But no one yet has been able to show that it actually has the ability to overcome this universal tendency, and that is the basic reason why there is still no bona fide proof of evolution, past or present.

From the statements of evolutionists themselves, therefore, we have learned that there is no real scientific evidence for real evolution. The only observable evidence is that of very limited horizontal (or downward) changes within strict limits. Evolution never occurred in the past, is not occurring at present, and could never happen at all. . . .

REFERENCES CITED

Johnson, Norman A. 2000. Design Flaw. *American Scientist* 88 (May/June): 274.

Lieb, E. H., and Jakob Yngvason. 2000. A Fresh Look at Entropy and the Second Law of Thermodynamics," *Physics Today* 53 (April): 32.

Selection excerpted from:

Morris, Henry M. 2001. The Scientific Case Against Evolution: A Summary. Part II. *Impact* 331: 1–4.

• •

THE SECOND LAW OF THERMODYNAMICS PRECLUDES EVOLUTION: OPPOSING VIEWS

The Creative Trinity

. . . According to this creationist concept, a system can become entropy deficient only if three conditions are satisfied. (1) Free energy must be supplied to the system. This is actually incorrect, since a loss of energy can also generate an entropy deficiency; however, the need for the system to be open is universally recognized, so further discussion is unnecessary. (2) The system must contain an energy conversion mechanism. When creationists are pressed, we find that just about anything qualifies as having a "mechanism," including matter itself, so the statement becomes quite meaningless. (3) The system must contain a directing program. This is variously referred to as intelligence, information, control system, and so forth by creationists. The idea is that this directing program did not arise through natural processes but was created by God. The Creative Trinity can also be interpreted as a statement to the effect that there are different kinds of entropy which are not interchangeable.

We must take careful note of an elementary fact which is often missed in debates on evolution and the second law: In spite of what they claim, creationists are no longer talking about the second law. They wish to give the impression that science, in this case thermodynamics, is on their side in their opposition to evolution. But the fact is there is nothing in thermodynamics that contradicts the phenomenon of an entropy deficiency being produced in a system when energy flows through it. On the contrary, this is what thermodynamics leads us to expect, and nothing else is needed, such as a directing program, etc. . . .

Creationists are not showing that evolution contradicts the second law of thermodynamics; instead, they are saying that the second law, as accepted by conventional science, is incorrect and insufficient to explain natural phenomena. They insist that something else of their own making must be added—namely, a divinely created directing program or a distinction between different kinds of entropy (p. 11).

Selection excerpted from:

Freske, Stanley. 1981. Creationist Misunderstanding, Misrepresentation, and Misuse of the Second Law of Thermodynamics. *Creation/Evolution* 2 (2): 8–16.

Biological Evolution and the Second Law

. . . Consider how different the world would be if all systems became less energetic and less organized with time. There would be no puffy clouds, thunderstorms, or weather fronts. Their organization and energy would have dissipated long ago. There would be no trees or flowers. Their seeds would just decay. And we wouldn't be here either. Each of us would have died as a withering zygote that could not undergo development. Clearly the creationist implication that all systems tend toward decay and disorder is wrong. There are many systems besides evolution that tend toward greater order. Philip Morrison (1978), for example, has shown that spontaneous increases in order are common in our world. He points out that the second law really says that increases in order must be paid for in energy. Such increases are clearly not impossible except in closed systems lacking a source of energy. Where large amounts of energy are available, as in the sun-earth system, large increases in order are possible.

Creationists, of course, deny this while claiming that organisms contain some sort of God-given precoded plan and energy conversion system that allows them to escape the death and decay dictated by the second law. On the other hand, almost all scientists accept both the second law and evolution. We need to ask, therefore, just how the second law does affect living systems. A look at gene mutation should allow an answer to this question. A given normal gene will mutate to a nonfunctional version of itself with a characteristic frequency, often on the order of 1/1,000,000. (For every 999,999 times this gene is transmitted correctly to the next generation, it is transmitted incorrectly one time.) We could call this type of mutation from functional to nonfunctional a "damaging" mutation.

It comes as a surprise to some people, but nonfunctional genes occasionally mutate back to the functional version. We could call this a "repair" mutation. If genes were likened to cars, this would be like saying that occasionally a dented car could be correctly fixed by being in a second accident! However, genes are not cars; chemical complexity is not the same thing as physical complexity. Even though an explosion in a print shop will not produce a dictionary, energy can change simple methane and ammonia into complex amino acids, as Stanley Miller and Harold Urey demonstrated in 1953. Similarly, even though a second collision probably will not undent a dented car, a second mutational event occasionally renders a gene functional again.

The effect of the second law is clearly seen when the repair mutation rate is measured. This repair rate is always less than the damaging mutation rate. In other words, it is easier to go from an ordered state (functional) to a disordered state (nonfunctional) than it is to go in the reverse direction. A typical rate for this repair type of mutation is on the order of 1/1,000,000,000. This is the most important consequence of the second law on living systems. Clearly, the second law does not prevent systems from going from disorder to order. All the law does in this case is to make such mutations rare compared to mutations going in the thermodynamically favored direction—toward disorder. If that's all there were to it, however, gene systems would still eventually all move to a disordered nonfunctional state. They obviously don't. Is this because of a mystical precoded plan, or is there another, nonsupernatural explanation?

Now we come to the essence of evolution: natural selection. All that any organism has to do to escape "degeneration in accord with the second law of thermodynamics" is to be able to produce more young than are needed to replace the parents. As long as that is true, the occasional mutants (almost all less fit than the original version) will usually reproduce poorly or even die without adversely affecting the population. Since the harmful mutations are underrepresented in succeeding generations, these mutations simply cannot build up to a level that threatens the well-being of the population. Thus, mutations are random changes, usually toward disorder, but the effect of natural selection is to remove the relatively common disordered genes and prevent the genetic system from degenerating.

In the same way, natural selection can replace genes with the rare mutant genes that represent an improvement over the original, thus serving as a type of ratchet to improve the organism and keep it matched to its changing environment. The entropy cost of the second law is paid as the energy required to produce those individuals that did not survive. The net result is that life opportunistically saves, builds upon, and improves whatever will function. At first glance, this may appear to conflict with the second law of thermodynamics, but the apparent conflict is not real. Therefore, no divinely precoded plan or mystical "vital force" is needed. Life and evolution are natural phenomena.

REFERENCES CITED

Morrison, P. 1978. In *On Aesthetics in Science*, ed. J. Weehsleo, p. 69. Cambridge, MA: MIT Press.

Selection excerpted from:

Thwaites, William, and Frank Awbrey. 1981. Biological Evolution and the Second Law. *Creation/Evolution* 2 (2) Issue 4: 5–7.

• •

THE VAPOR CANOPY THEORY: SELECTION SUPPORTING CREATIONIST VIEWS

The Sky Has Fallen

Introduction

The world-wide flood recounted in Genesis has no parallel in today's world. Yet, few serious attempts have been made in the past to explore the meteorology of the flood and the atmosphere of the antediluvian world. Several advances have recently been made in developing atmospheric models and comparing model predictions with observations. These attempts to understand what the atmosphere (firmament) was like before and during the flood help us to realize that, indeed, "The Sky Has Fallen."

Models

. . . The conceptual vapor canopy model developed by Dillow specifies that the earth was surrounded by a vapor canopy before the flood of Noah. This pre-flood atmosphere contained the equivalent of about 40 feet of water in the form of a canopy resting on top of the current atmosphere. The canopy condensed suddenly during the 40-day period of Noah's flood, causing the universal deluge. Given such a conceptual model, at least three predictions can be compared with appropriate observations to help confirm or refute the model.

1. An extensive greenhouse effect would have occurred prior to the flood.
2. Physical processes would have been different and plant and animal life would have been affected by the increased atmospheric pressure under the vapor canopy.
3. Temperatures in the polar regions would have decreased suddenly and permanently.

Greenhouse Effect. The greenhouse effect gets its name from the observation that the air inside a greenhouse is warmer than the air outside because heat is trapped by the glass windows. Shortwave radiation from the sun travels relatively unimpeded through the glass, but longwave radiation returning from the plants and earth inside the greenhouse cannot easily be transmitted back through the glass. Consequently, the heat is trapped and the temperature in the greenhouse rises. A similar effect occurs in our atmosphere today. If it were not for this effect the surface of the earth would be like the moon, which gets extremely hot during the day and extremely cold at night.

Prior to the flood the greenhouse effect would have been amplified greatly. An amplified greenhouse effect would have not only caused the atmosphere to be warmer but would have tended to create a uniform temperature distribution from equator to poles. In addition, it is likely that the temperature in the canopy would have been greater than that near the surface of the earth. In the pre-flood atmosphere, if one were to have gone to the mountains to cool off, assuming there were any mountains prior to the flood, he would have found that the temperature increased rather than decreased as he got higher. Such a condition is called an inversion. We know that such conditions lead to pollution episodes around large cities today because under an inversion the air is very stable and the winds are very light to non-existent. In the pre-flood atmosphere the inversion would have been very strong and the pole-to-equator temperature difference would have been very small, resulting in light winds, no storms, and no rain! The entire earth, including the poles, would have been much warmer than it is today.

There is abundant evidence that the polar regions were much warmer at one time. A fallen 90-foot fruit tree with ripe fruit and green leaves still on its branches has been found in the frozen ground of the New Siberian islands. The only tree vegetation that grows there now is the one-inch-high willow. Palm tree fossils have been found in early Tertiary strata in Alaska. Large fossil leaves of tropical plants have been found in Permian sandstone 250 miles from the South Pole. Crocodiles were once prolific in New Jersey and England. It is estimated that the mean sea-level air temperature at the poles was 45 degrees Fahrenheit during the Cretaceous period. Today, the temperature is –4 degrees Fahrenheit. The evidence of warm polar regions is so extensive that the theory of continental drift was developed by evolutionary geologists to help explain how tropical fossil material can be accounted for at the poles. The vapor canopy theory, on the other hand, explicitly predicts tropical vegetation at the poles without the need for refinements to the theory.

Increased Atmospheric Pressure. Pressure is the weight pressing on a surface per unit area. Pressure increases the lower one goes in the atmosphere because there is more mass of vapor stacked above. Prior to the flood, when the vapor canopy was resting on top of the ancient atmosphere, its additional weight would have approximately doubled the surface pressure we experience today.

There are several features in the geologic record which might be explained by greater atmospheric pressure at some time in the past. One of the puzzles of natural history is the gigantic flying reptile called the pteranodon. This flying reptile had wingspans of up to 20 feet. Many authors have questioned how such an animal could launch itself into the air from flat ground. The minimum speed for the pteranodon has been computed to be more than 15 mph in today's atmosphere. Since the pteranodon could not run, this meant that a wind of more than 15 mph would have had to occur before the reptile could become airborne. Pilots know, however, that it is easier to take off at lower altitudes, where the pressure is greater. If the atmospheric pressure were twice what it is today prior to the flood, it would have been much easier and required much lighter winds for the pteranodon to take off. Calculations show that it would have required a wind of just over 10 mph for the pteranodon to get airborne in the pre-flood atmosphere.

. . . Another illustration of the possible effects of greater atmospheric pressure before the flood is the presence of gigantism in the fossil record. Giant dinosaurs weighing over 40 tons, insects with 25-inch wingspans, and giant shell creatures, spiders, and other invertebrates once lived on the earth, but not today. Is it possible that the greater pressure in the pre-flood atmosphere was able to help supply more oxygen to the biomass of these animals, allowing them to live longer, healthier lives and grow larger?

Evidence that higher oxygen pressures are beneficial to biological systems was recently discovered in the aquanaut program. One of the aquanauts reported that a severe cut on his hand healed completely within 24 hours while submerged in a diving bell at a pressure of 10 atmospheres. It was theorized that the higher pressure forced more oxygen into the tissue surrounding the wound and healed it at a greater rate. Based on this observation, experiments in hyperbaric surgery were started with excellent results. Higher atmosphere pressure has been found to result in relief from some effects of aging and the cure of some other diseases. It is not hard to believe that such an effect could be related in some way to gigantism and the longevity of life evident before the flood.

Polar Temperature Decrease. With the condensation and collapse of the vapor canopy, the warm climate it produced likely disappeared suddenly over the 40-day period of the flood. The radiation balance at the poles is such that without a canopy the temperature would rapidly drop below freezing. Animals caught in the flood, cold, and wind would be frozen rapidly along with the sediment from the flood.

The bones of thousands of animals have, in fact, been found frozen in the tundra of Siberia. Hippopotamuses, saber-tooth tigers, mammoths, and other animals normally associated with the tropics have been found frozen, some in relatively fresh condition, in the frozen Siberian muck. This muck is full of plant and animal remains to depths of several thousand feet. . . .

Conclusions

Such a controversial model is bound to create discussion and criticism. At the same time, however, it will increase the interest and enthusiasm of specialists in the atmospheric sciences and the canopy theory. More quantification of such mathematical models is desirable and will result in further improvements of our understanding of the flood and the antecedent atmosphere. The final result will produce even greater confidence in the Word of God.

Selection excerpted from:

Vardiman, Larry. 1984. The Sky Has Fallen. *Impact* 128: 1–4.

• •

FLOOD GEOLOGY: SELECTIONS REFUTING CREATIONIST VIEWS

The Water Canopy Theory

The canopy theory or model requires that there existed in the golden age following creation, but preceding the Flood, an envelope of water [vapor] completely surrounding the earth in a zone suspended between the earth's surface and the ozone layer. . . . The vapor canopy . . . is postulated to have contained the equivalent of a water layer about 12 m (40 ft) in depth over the entire earth. It is described as "resting on top of the current atmosphere" (Vardiman 1984: 1). The Flood hypothesis calls for a rapid condensation of the vapor in the canopy over a 40-day period, causing it to fall to earth as the Flood, or universal Deluge. . . (p. 195).

Creationists have made the vapor canopy theory a kingbolt of the Flood scenario. But does it "hold water" in terms of what is known of atmospheric science? I am concerned about the stability of such a system. It seems to me that the canopy could not exist for two reasons. First, water vapor, if not isolated or enclosed by an impervious barrier, quickly diffuses into surrounding regions of lower vapor density. Everyone who uses a vaporizer (humidifier) in the home or who boils a pot of water on the stove in a closed room knows this. Water vapor of a dense "canopy"—should such a canopy have momentarily come into existence for reasons unspecified—would quickly diffuse downward and upward, until it attained a uniform distribution over the globe such that the proportion of water molecules would be in a uniform ratio to the other molecules of the atmosphere (largely nitrogen and oxygen). In other words, the water vapor would respond to the laws of behavior of gaseous molecules. The density of the gaseous mixture would quickly reach a distribution like that found today—densest at the surface and diminishing upward exponentially to the upper limit of the atmosphere . . . (p. 195).

The creation scientists have, in their canopy model, used a water-equivalent depth of 12 m (Vardiman 1984: 1). Just how much more water may have come from the subterranean sources is not mentioned, but a uniform addition of 12 m to existing sea level would not be much of a flood—certainly not if it were to land the Ark on Mount Ararat. Two professors of earth science at St. Cloud State University, Minnesota, Leonard Soroka and Charles L. Nelson, have made extensive calculations of the total quantity of water required to raise present sea level to the height of Mount Everest, so as to satisfy the Genesis provision of covering all mountaintops (Soroka and Nelson 1983). The total requirement comes to 4.4 billion cubic kilometers. They conclude that to accommodate this volume as water vapor in the atmosphere is impossible. They give four reasons for that conclusion:

> First, atmospheric pressure would be about 840 times higher than it is now. Second, the atmosphere would be 99.9 percent water vapor and it would be impossible for humans and other animals to breathe such an atmosphere. Third, such a mixture of gases could not even exist as gases at temperatures that humans could tolerate. Fourth, neither could the water be accommodated as cloud droplets since there would be insufficient nitrogen and oxygen (less than .1 percent) to support them and since as clouds they would have prevented nearly all sunlight from reaching the surface. In short, such an atmosphere would not have allowed terrestrial life as we know it to exist on the surface of the earth. (Soroka and Nelson 1983: 135)

Even more damaging to the theory of an atmospheric water source of such magnitude is the thermodynamic effect of condensation of the water vapor. Soroka and Nelson show that the rise in temperature caused by liberation of the latent heat of vaporization during a 40-day period would raise the atmospheric temperatures over the entire earth to over 3,500 C (6,400 F). Before this heat could have been dissipated by radiation into space, it would have set the ocean to boiling and would have cremated the Ark (p. 135). The same authors also evaluate the subterranean water source through springs, assuming that the subterranean source must have supplied most of the 4.4 billion cubic kilometers. If that much water were to be exuded from the crust of the continents and oceans, the porosity of the crust must be at least 50 percent (half solid mineral matter and half open voids). Actual porosity of crustal rock is much less than 1 percent because rock under the pressure that exists at depths of several kilometers is ductile in behavior and capable of closing any open pores that might exist (p. 197).

REFERENCES CITED

Soroka, Leonard G., and Charles L. Nelson. 1983. Physical Constraints on the Noachian Deluge. Journal of Geological Education 31: 135–139.

Vardiman, Larry. 1984. The Sky Has Fallen. *Impact* 128: 1–4.

Selection excerpted from:

Strahler, Arthur. 1987. *Science and Earth History: The Evolution/Creation Controversy*. Buffalo, NY: Prometheus Books.

RADIOMETRIC DATING: SELECTIONS SUPPORTING CREATIONIST VIEWS

Radiometric Dating

In attempting to determine the real age of the earth, it should always be remembered, of course, that recorded history began only several thousand years ago. Not even uranium dating is capable of experimental verification, since no one would actually watch uranium decaying for millions of years to see what happens.

In order to obtain a prehistoric date, therefore, it is necessary to use some kind of physical process which operates slowly enough to measure and steadily enough to produce significant changes. If certain assumptions are made about it, then it can yield a date which could be called the *apparent age*. Whether or not the apparent age is the *true age* depends completely on the validity of the assumptions. Since there is no way in which the assumption can be tested, there is no *sure* way (except by divine revelation) of knowing the true age of any geologic formation. The processes which are most likely to yield dates, which approximate the true dates, are those for which the assumptions are least likely to be in error. . . .

As far as the age of geological formations and of the earth itself are concerned, only radioactive decay processes are considered useful today by evolutionists. There are a number of these, but the most important ones are: (1) the various uranium-thorium-lead methods; (2) the rubidium-strontium method, and (3) the potassium-argon method. In each of these systems, the parent (e.g., uranium) is gradually changed to the daughter (e.g., lead) component of the system, and the relative proportions of the two are considered to be an index of the time since initial formation of the system.

For these or other methods of geochronometry, one should note carefully that the following assumptions must be made:

(1) *The system must have been a closed system.*
That is, it cannot have been altered by factors extraneous to the dating process; nothing inside the system could have been removed, and nothing outside the system added to it.
(2) *The system must initially have contained none of its daughter component.*
If any of the daughter component were present initially, the initial amount must be corrected in order to get a meaningful calculation.
(3) *The process rate must always have been the same.*

Similarly, if the process rate has ever changed since the system was established, then this change must be known and corrected for if the age calculation is to be of any significance.

Other assumptions may be involved for particular methods, but the three listed above are always involved and are critically important. In view of this fact, the highly speculative nature of all methods of geochronometry becomes apparent when one realizes that *not one* of the above assumptions is valid! None are provable, or testable, or even reasonable.

Selection excerpted from:

Morris, Henry M., ed. 1974. *Scientific Creationism.* San Diego: Creation-Life Publishers.

What About Carbon Dating?

Other Radiometric Dating Methods

There are various other radiometric dating methods used today to give ages of millions or billions of years for rocks. These techniques, unlike carbon dating, mostly use the relative concentrations of parent and daughter products in radioactive decay chains. For example, potassium-40 decays to argon-40; uranium-238 decays to lead-206 via other elements like radium; uranium-235 decays to lead-207; rubidium-87 decays to strontium-87; etc. These techniques are applied to igneous rocks, and are normally seen as giving the time since solidification.

The isotope concentrations can be measured very accurately, but isotope concentrations are not dates. To derive ages from such measurements, unprovable assumptions have to be made such as:

1. The starting conditions are known (for example, that there was no daughter isotope present at the start, or that we know how much was there).
2. Decay rates have always been constant.
3. Systems were closed or isolated so that no parent or daughter isotopes were lost or added. . . .

"Bad" Dates

When a "date" differs from that expected, researchers readily invent excuses for rejecting the result. The common application of such posterior reasoning shows that radiometric dating has serious problems. Woodmorappe (1999) cites hundreds of examples of excuses used to explain "bad" dates.

For example, researchers applied posterior reasoning to the dating of *Australopithecus ramidus* fossils (WoldeGabriel et al. 1994). Most samples of basalt closest to the fossil-bearing strata give dates of about 23 Ma (Mega annum, million years) by the argon-argon method. The authors decided that was "too old," according to their beliefs about the place of the fossils in the evolutionary grand scheme of things. So they looked at some basalt further removed from the fossils and selected 17 of 26 samples to get an acceptable maximum age of 4.4 Ma. The other nine samples again gave much older dates but the authors decided they must be contaminated and discarded them. That is how radiometric dating works. It is very much driven by the existing long-age world view that pervades academia today.

A similar story surrounds the dating of the primate skull known as KNM-ER 1470. This started with an initial 212 to 230 Ma, which, *according to the fossils*, was considered way off the mark (humans "weren't around then"). Various other attempts were made to date the volcanic rocks in the area. Over the years an age of 2.9 Ma was settled upon because of the agreement between several different published studies (although the studies involved selection of "good" from "bad" results, just like *Australopithecus ramidus*, above).

However, preconceived notions about human evolution could not cope with a skull like 1470 being "that old." A study of pig fossils in Africa readily convinced most anthropologists that the 1470 skull was much younger. After this was widely accepted, further studies of the rocks brought the radiometric age down to about 1.9 Ma—again several studies "confirmed" *this* date. Such is the dating game.

Are we suggesting that evolutionists are conspiring to massage the data to get what they want? No, not generally. It is simply that all observations must fit the prevailing paradigm. The paradigm, or belief system, of molecules-to-man evolution over eons of time, is so strongly entrenched it is not questioned—it is a "fact." So every observation must fit this paradigm. Unconsciously, the researchers, who are supposedly "objective scientists" in the eyes of the public, select the observations to fit the basic belief system.

We must remember that the past is not open to the normal processes of experimental science, that is, repeatable experiments in the present. A scientist cannot do experiments on events that happened in the past. Scientists do not measure the age of rocks, they measure isotope concentrations, and these can be measured extremely accurately. However, the "age" is calculated using assumptions about the past that cannot be proven.

We should remember God's admonition to Job, "Where were you when I laid the foundations of the earth?" (Job 38:4).

Those involved with unrecorded history gather information in the present and construct stories about the past. The level of proof demanded for such stories seems to be much less than for studies in the empirical sciences, such as physics, chemistry, molecular biology, physiology, etc.

Williams, an expert in the environmental fate of radioactive elements, identified 17 flaws in the isotope dating reported in just three widely respected seminal papers that supposedly established the age of the earth at 4.6 billion years (Williams 1992). John Woodmorappe has produced an incisive critique of these dating methods (Woodmorappe 1999). He exposes hundreds of myths that have grown up around the techniques. He shows that the few "good" dates left after the "bad" dates are filtered out could easily be explained as fortunate coincidences.

What Date Would You Like?

The forms issued by radioisotope laboratories for submission with samples to be dated commonly ask how old the sample is expected to be. Why? If the techniques were absolutely objective and reliable, such information would not be necessary. Presumably, the laboratories know that anomalous dates are common, so they need some check on whether they have obtained a "good" date.

Testing Radiometric Dating Methods

If the long-age dating techniques were really objective means of finding the ages of rocks, they should work in situations where we know the age. Furthermore, different techniques should consistently agree with one another.

Methods Should Work Reliably on Things of Known Age

There are many examples where the dating methods give "dates" that are wrong for rocks of known age. One example is K-Ar "dating" of five historical andesite lava flows from Mount Nguaruhoe in New Zealand. Although one lava flow occurred in 1949, three in 1954, and one in 1975, the "dates" range from less than 0.27 to 3.5 Ma (Snelling 1998).

Again, using hindsight, it is argued that "excess" argon from the magma (molten rock) was retained in the rock when it solidified. The secular scientific literature lists many examples of excess argon causing dates of millions of years in rocks of known historical age (Snelling 1998). This excess appears to have come from the upper mantle, below the earth's crust. This is consistent with a young world—the argon has had too little time to escape (Snelling 1998). If excess argon can cause exaggerated dates for rocks of known age, then why should we trust the method for rocks of unknown age?

Other techniques, such as the use of isochrons, make different assumptions about starting conditions, but there is a growing recognition that such "foolproof" techniques can also give "bad" dates. So data are again selected according to what the researcher already believes about the age of the rock.

Geologist Dr. Steve Austin sampled basalt from the base of the Grand Canyon strata and from the lava that spilled over the edge of the canyon. By evolutionary reckoning, the latter should be a billion years younger than the basalt from the bottom. Standard laboratories analyzed the isotopes. The rubidium-strontium isochron technique suggested that the recent lava flow was 270 Ma older than the basalts beneath the Grand Canyon—an impossibility.

Different Dating Techniques Should Consistently Agree

If the dating methods are an objective and reliable means of determining ages, they should agree. If a chemist were measuring the sugar content of blood, all valid methods for the determination would give the same answer (within the limits of experimental error). However, with radiometric dating, the different techniques often give quite different results.

In the study of the Grand Canyon rocks by Austin, different techniques gave different results (Austin 1994: 84–85). Again, all sorts of reasons can be suggested for the "bad" dates, but this is again posterior reasoning. Techniques that give results that can be dismissed just because they don't agree with what we already believe cannot be considered objective.

Method	"Age"
Six potassium-argon model ages	10,000 years to 117 Ma
Five rubidium-strontium ages	1,270–1,390 Ma
Rubidium-strontium isochron	1,340 Ma
Lead-lead isochron	2,600 Ma

REFERENCES CITED (more available in original article)

Austin, S.A., ed. 1994. *Grand Canyon: Monument to Catastrophe*, pp. 120–123. Santee, CA: Institute for Creation Research.

Snelling, A. A. 1998. The Cause of Anomalous Potassium-Argon "Ages" for Recent Andesite Flows at Mt. Nguaruhoe, New Zealand, and the Implications for Potassium-Argon "Dating." In *Proc. 4th ICC*, pp. 503–525.

Williams, A. R. 1992. Long-age Isotope Dating Short on Credibility. *CEN Technical Journal* 6 (1): 2–5.

WoldeGabriel, G., et al. 1994. Ecological and Temporal Placement of Early Pliocene Hominids at Aramis, Ethiopia. *Nature* 371: 330–333.

Woodmorappe, J. 1999. *The Mythology of Modern Dating Methods*. San Diego: Institute for Creation Research.

Selection excerpted from:

Batten, Don, ed., Ken Ham, Jonathan D. Sarfati, and Carl Wieland. 2000. *The Revised and Expanded Answers Book: The 20 Most-Asked Questions About Creation, Evolution and the Book of Genesis Answered!*, rev. ed. Green Forest, AR: Master Books.

●●●●●●●●●●●●●●●●●●●●●●●●●●●●●●

RADIOMETRIC DATING: SELECTION REFUTING CREATIONIST VIEWS

Common Misconceptions Regarding Radiometric Dating Methods

1. Radiometric dating is based on index fossils whose dates were assigned long before radioactivity was discovered.

This is not at all true, though it is implied by some young-Earth literature. Radiometric dating is based on the half-lives of the radioactive isotopes. These half-lives have been measured over the last 40–90 years. They are not calibrated by fossils.

2. No one has measured the decay rates directly; we only know them from inference.

Decay rates have been directly measured over the last 40–100 years. In some cases a batch of the pure parent material is weighed and then set aside for a long time and then the resulting daughter material is weighed. In many cases it is easier to detect radioactive decays by the energy burst that each decay gives off. For this a batch of the pure parent material is carefully weighed and then put in front of a Geiger counter or gamma-ray detector. These instruments count the number of decays over a long time.

3. If the half-lives are billions of years, it is impossible to determine them from measuring over just a few years or decades.

The example given in the section titled "The Radiometric Clocks" shows that an accurate determination of the half-life is easily achieved by direct counting of decays over a decade or shorter. This is because a) all decay curves have exactly the same shape [figure omitted], differing only in the half-life, and b) trillions of decays can be counted in one year even using only a fraction of a gram of material with a half-life of a billion years. Additionally, lavas of historically known ages have been correctly dated even using methods with long half-lives.

4. The decay rates are poorly known, so the dates are inaccurate.

Most of the decay rates used for dating rocks are known to within two percent. Uncertainties are only slightly higher for rhenium (5%), lutetium (3%), and beryllium (3%), discussed in connection with Table 1 [table omitted]. Such small uncertainties are no reason to dismiss radiometric dating. Whether a rock is 100 million years or 102 million years old does not make a great deal of difference. . . .

6. Decay rates can be affected by the physical surroundings.

This is not true in the context of dating rocks. Radioactive atoms used for dating have been subjected to extremes of heat, cold, pressure, vacuum, acceleration, and strong chemical reactions far beyond anything experienced by rocks, without any significant change. The only exceptions, which are not relevant to dating rocks, are discussed under the section "Doubters Still Try," above. . . .

10. To date a rock one must know the original amount of the parent element. But there is no way to measure how much parent element was originally there.

It is very easy to calculate the original parent abundance, but that information is not needed to date the rock. All of the dating schemes work from knowing the present abundances of the parent and daughter isotopes. The original abundance N_0, of the parent, is simply $N_0 = Ne^{kt}$, where N is the present abundance, t is time, and k is a constant related to the half-life.

11. There is little or no way to tell how much of the decay product, that is, the daughter isotope, was originally in the rock, leading to anomalously old ages.

A good part of this article is devoted to explaining how one can tell how much of a given element or isotope was originally present. Usually it involves using more than one sample from a given rock. It is done by comparing the ratios of parent and daughter isotopes relative to a stable isotope for samples with different relative amounts of the parent isotope. For example, in the rubidium-strontium method one compares rubidium-87/strontium-86 to strontium-87/strontium-86 for different minerals. From this one can determine how much of the daughter isotope would be present if there had been no parent isotope. This is the same as the initial amount (it would not change if there were no parent isotope to decay). Figures 4 and 5 [omitted], and the accompanying explanation, tell how this is done most of the time. While this is not absolutely 100% foolproof, comparison of several dating methods will always show whether the given date is reliable.

12. There are only a few different dating methods.

This article has listed and discussed a number of different radiometric dating methods and has also briefly described a number of non-radiometric dating methods. There are actually many more methods out there. Well over forty different radiometric dating methods are in use, and a number of non-radiogenic methods not even mentioned here. . . .

14. A young-Earth research group reported that they sent a rock erupted in 1980 from Mount Saint Helens volcano to a dating lab and got back a potassium-argon age of several million years. This shows we should not trust radiometric dating.

There are indeed ways to "trick" radiometric dating if a single dating method is improperly used on a sample. Anyone can move the hands on a clock and get the wrong time. Likewise, people actively looking for incorrect radiometric dates can in fact get them. Geologists have known for over forty years that the potassium-argon method cannot be used on rocks only twenty to thirty years old. Publicizing this incorrect age as a completely new finding was inappropriate. The reasons are discussed in the Potassium-Argon Dating section above. Be assured that multiple dating methods used together on igneous rocks are almost always correct unless the sample is too difficult to date due to factors such as metamorphism or a large fraction of xenoliths. . . .

18. We know the Earth is much younger because of non-radiogenic indicators such as the sedimentation rate of the oceans.

There are a number of parameters which, if extrapolated from the present without taking into account the changes in the Earth over time, would seem to suggest a somewhat younger Earth. These arguments can sound good on a very simple level, but do not hold water when all the factors are considered. Some examples of these categories are the decaying magnetic field (not mentioning the widespread evidence for magnetic reversals), the saltiness of the oceans (not counting sedimentation!), the sedimentation rate of the oceans (not counting Earthquakes and crustal movement, that is, plate tectonics), the relative paucity of meteorites on the Earth's surface (not counting weathering or plate tectonics), the thickness of dust on the moon (without taking into account brecciation over time), the Earth-Moon separation rate (not counting changes in tides and internal forces), etc. While these arguments do not stand up when the complete picture is considered, the case for a very old creation of the Earth fits well in all areas considered. . . .

20. Different dating techniques usually give conflicting results.

This is not true at all. The fact that dating techniques most often agree with each other is why scientists tend to trust them in the first place. Nearly every college and university library in the country has periodicals such as *Science*, *Nature*, and specific geology journals that give the results of dating studies. The public is usually welcome to (and should!) browse in these libraries. So the results are not hidden; people can go look at the results for themselves. Over a thousand research papers are published a year on radiometric dating, essentially all in agreement.

[Figures and tables omitted.]

Selection excerpted from:

Weins, Roger C. 2003. *Radiometric Dating: A Christian Perspective*. ASA Resources, 2002. Accessed July 26, 2003. Available from http://www.asa3.org/ASA/resources/Wiens.html#page%2023.

• •

THE ANTHROPIC PRINCIPLE: SELECTION SUPPORTING CREATIONIST VIEWS

The Harmony of the Spheres

Physicists recognize four fundamental forces. These largely determine the way in which one bit of matter or radiation can interact with another. In effect, these four forces determine the main characteristics of the universe (Trimble 1977). They are the gravitational force, the electromagnetic force, the strong or nuclear force, and the weak force.

An extraordinary feature of these four fundamental forces is that their strength varies enormously over many orders of magnitude. In the table below they are given in international standard units (Boslough 1985).

The forces of nature

Gravitational force	5.9–10^{-39}
Nuclear or strong force	15
Electromagnetic force	3.05–10^{-12}
Weak force	7.03–10^{-3}

The fact that the gravitational force is fantastically weaker than the strong nuclear force by an unimaginable thirty-eight orders of magnitude is critical to the whole cosmic scheme and particularly to the existence of stable stars and planetary systems (Boslough, 1985). If, for example, the gravitational force was a trillion times stronger, then the universe would be far smaller and its life history far shorter. An average star would have a mass a trillion times less than the sun and a life span of about one year—far too short a time for complex life to develop and flourish. On the other hand, if gravity had been less powerful, no stars of galaxies would ever have formed. As Hawking points out, the growth of the universe—so close to the border of collapse and external expansion that man has not been able to measure it—has been at just the proper rate to allow galaxies and stars to form (Boslough 1985).

The other relationships are no less critical. If the strong force had been just slightly weaker, the only element that would be stable would be hydrogen. No other atoms could exist. If it had been slightly stronger in relation to electromagnetism, then an atomic nucleus consisting of only two protons would be a stable feature of the universe—which would mean there would be no hydrogen, and if any stars or galaxies evolved, they would be very different from the way they are (Gribben and Rees 1989).

Clearly, if these various forces and constants did not have precisely the values they do, there would be no stars, no supernovae, no planets, no atoms, no life. . . .

In short, the laws of physics are supremely fit for life and the cosmos gives every appearance of having been specifically and optimally tailored to that end: to ensure the generation of stable stars and planetary systems; to ensure that these will be far enough apart to avoid gravitational interactions which would destabilize planetary orbits; to ensure that a nuclear furnace is generated in the interior of stars in which hydrogen will be converted into the heavier elements essential for life; to ensure that a proportion of stars will undergo supernovae explosions to release the key elements into interstellar space; to ensure that galaxies last several times longer than the lifetime of an average star, for only then will there be time for the atoms scattered by an earlier generation of supernovae within any one galaxy to be gathered into second-generation solar systems; to ensure that the distribution and frequency of supernovae will not be so frequent that planetary surfaces would be repeatedly bathed in lethal radiation but not so infrequent that there would be no heavier atoms manufactured and gathered on the surface of newly formed planets; to ensure in the cosmos's vastness and in the trillions of its suns and their accompanying planetary systems a stage immense enough and a time long enough to make certain that the great evolutionary drama of life's becoming will inevitably be manifest sometime, somewhere on an earthlike planet.

And so we are led toward life and our own existence via a vast and ever-lengthening chain of apparently biocentric adaptations in the design of the cosmos in which each adaptation seems adjusted with almost infinite precision toward the goal of life (pp. 12–14).

REFERENCES CITED

Boslough, J. 1985. *Stephen Hawking's Universe*, p. 101. New York: Quill.

Gribben, J. R., and M. J. Rees. 1989. *Cosmic Coincidences*, chap. 10, pp. 241–269. New York: Bantam Books.

Trimble, V. 1977. Cosmology: Man's Place in the Universe. *American Scientist* 65: 76–86.

Selection excerpted from:

Denton, Michael. 1998. *Nature's Destiny: How the Laws of Biology Reveal Purpose in the Universe*. New York: Free Press.

• •

THE ANTHROPIC PRINCIPLE: SELECTION REFUTING CREATIONIST VIEWS

Anthropic Design: Does the Universe Show Evidence of Purpose?

. . . The fine-tuning argument is based on the fact that earthly life is very sensitive to the values of several fundamental physical constants. Making the tiniest change in any of these, and life as we know it would not exist. The delicate connections between physical constants and life are called the anthropic coincidences (Carter 1974; Barrow and Tipler 1986). The name is a misnomer. Human life is not singled out in any special way. At most, the coincidences show that the production of carbon and the other elements that make earthly life possible required a sensitive balance of physical parameters. . . (p. 41).

The interpretation of the anthropic coincidences in terms of purposeful design should be recognized as yet another variant of the ancient argument from design that has appeared in many different forms over the ages. The anthropic design argument asks: how can the universe possibly have obtained the unique set of physical constants it has, so exquisitely fine-tuned for life as they are, except by purposeful design—design with life and perhaps humanity in mind?

This argument, however, has at least one fatal flaw. It makes the wholly unwarranted assumption that only one type of life is possible—the particular form of carbon-based life we have here on earth. Even if this is an unlikely result of chance, some form of life could still be a likely result. It is like arguing that a particular card hand is so improbable that it must have been foreordained.

Based on recent studies in the sciences of complexity and "Artificial Life" computer simulations, sufficient complexity and long life appear to be primary conditions for a universe to contain some form of reproducing, evolving structures. This can happen with a wide range of physical parameters, as has been demonstrated (Stenger 1995). The fine-tuners have no basis in current knowledge for assuming that life is impossible except for a very narrow, improbable range of parameters . . . (pp. 41–42).

The inflationary big bang offers a plausible, natural scenario for the uncaused origin and evolution of the universe, including the formation of order and structure—without the violation of any laws of physics. These laws themselves are now understood far more deeply than before, and we are beginning to grasp how they too could have come about naturally . . . (p. 42).

So how did our universe happen to be so "fine-tuned" as to produce wonderful, self-important carbon structures . . . ? If we have no reason to assume ours is the only life form, we also have no reason to assume that ours is the only universe. Many universes can exist, with all possible combinations of physical laws and constants. In that case, we just happen to be in the particular one that was suited for the evolution of our form of life. When cosmologists refer to the anthropic principle, this is all they usually mean. Since we live in this universe, we can assume it possesses qualities suitable for our existence. Humans evolved eyes sensitive to the region of electromagnetic spectrum from red to violet because the atmosphere is transparent in that range. Yet some would have us think that the causal action was the opposite, that the atmosphere of the earth was designed to be transparent from red to violet because human eyes are sensitive in that range. Stronger versions of the anthropic principle, which assert that the universe is somehow actually required to produce intelligent "information-processing systems" (Barrow and Tipler 1986), are not taken seriously by most scientists or philosophers . . . (p. 42).

The existence of many universes is consistent with all we know about physics and cosmology (Smith 1990; Smolin 1992, 1997; Linde 1994; Tegmark 1997). Some theologians and scientists dismiss the notion as a gross violation of Occam's razor (see, for example, Swinburne 1990). It is not. No new hypothesis is needed to consider multiple universes. In fact, it takes an added hypothesis to rule them out—a super law of nature that says only one universe can exist. But we know of no such law, so we would violate Occam's razor to insist on only one universe.

REFERENCES CITED

Barrow, John D., and Frank J. Tipler. 1986. *The Anthropic Cosmological Principle*. Oxford: Oxford University Press.

Carter, Brandon. 1974. Large Number Coincidences and the Anthropic Principle in Cosmology. In M. S. Longair, ed., *Confrontation of Cosmological Theory with Astronomical Data*, pp. 291–298. Dordrecht, Netherlands: Reidel.

Leslie, John. 1990. *Physical Cosmology and Philosophy*, New York: Macmillan.

Linde, Andre. 1994. "The Self-Reproducing Inflationary Universe." *Scientific American*, November, 48–55.

Smith, Quentin. 1990. A Natural Explanation of the Existence and Laws of Our Universe. *Australasian Journal of Philosophy* 68: 22–43.

Smolin, Lee. 1992. Did the Universe Evolve? *Classical and Quantum Gravity* 9: 173–191.

Smolin, Lee. 1997. *The Life of the Cosmos*. Oxford: Oxford University Press.

Stenger, Victor J. 1995. *The Unconscious Quantum: Metaphysics in Modern Physics and Cosmology*. Amherst, NY: Prometheus Books.

Swinburne, Richard. 1990. "Argument from the Fine-Tuning of the Universe" in Leslie (1990: 154–173).

Tegmark, Max. 19987. Is "The Theory of Everything" Merely the Ultimate Ensemble Theory? *Annals of Physics*, 270: 1–51.

Selection excerpted from:

Stenger, Victor. 1999. Anthropic Design: Does the Universe Show Evidence of Purpose? *Skeptical Inquirer* 23 (4): 40–43.

CHAPTER 8

Patterns and Processes of Biological Evolution

INTRODUCTION

Biological evolution is uniformly rejected by Young Earth, Old Earth, and most Intelligent Design creationists. In this chapter we will look at how creationists and evolutionary biologists look at biological evolution's patterns (the relationships of living things through time) and processes (the forces or factors that are thought to bring about change). Three themes pertinent to the pattern of evolutionary biology are commonly encountered in antievolutionist literature: the presence of gaps in the fossil record, the Cambrian Explosion, and the question of design. Processes of evolution may conceptually be separated from the patterns of evolution, but as you will read, writers often discuss both of them together.

The key question of process to antievolutionists is whether natural selection and other microevolutionary processes can account for what they call "macroevolution." This is discussed later in the chapter under the heading "Micro/Macro."

GAPS IN THE FOSSIL RECORD

Evolutionary biologists and antievolutionists are united in one respect: both agree that there are gaps in the fossil record. The record of life as seen in stone does not present a smooth, intergrading continuum from earliest times until the present, nor is there a continuum between all living things. Darwin accounted for the fact that most living species do not grade into one another (although some do) because extinction removes intermediate forms. But Darwin himself expected that as the fossil record became better known (it was scarcely investigated in 1859), fossils showing links between groups—transitional fossils—would be discovered.

Creationists claim that there is a systematic lack of transitional fossils; Special Creation explains why there are no intermediates. Evolutionary biologists point to myriad intermediates—generally unknown or unaccepted by creationists, and are unperturbed by the presence of gaps when they occur.

Part of the difference between the two positions is conceptual and definitional: creationists and evolutionists define and understand evolution differently. Creationists view evolution as being progressive and ladderlike, as in the Great Chain of Being discussed in chapter 4. This and their understanding of natural selection predict a slow and gradual change of species through time which would result in the fossil record containing a finely graded succession of fossils, representing a finely graded succession of living things. Evolutionary biologists view the tree of life as resembling a bush more than a ladder, and recognize mechanisms in addition to natural selection, and that rates of change can be gradual or rapid. The casualness with which evolutionary biologists accept gaps also reflects their recognition that the fossil record represents only a fraction of a fraction of all the species that have ever lived; gaps are to be expected.

Gaps: The Creationist View

We had intended to excerpt a passage from the Intelligent Design textbook *Of Pandas and People* (1993), but the authors denied us permission. In the section of *Pandas . . .* titled "Fossil Stasis and Gaps Within the Phyla" (pp. 100–101), Percival W. Davis and Dean H. Kenyon begin with the claim that much of the fossil record is characterized by a lack of transitions: there is no continuous transitional series between, for example, the earliest horse and the contemporary horse or reptiles and mammals. Davis and Kenyon quote the paleontologists David Raup, Stephen Jay Gould, and Steven M. Stanley, and the morphologist Harold C. Bold, all writing in the 1960s or 1970s, for support. Turning to a specific example—the series from reptiles to mammals by way of a group of mammal-like reptiles called therapsids—they address a discussion by James Hopson, writing in *The American Biology Teacher*. Considering a series of eight therapsids and comparing them to the early mammal *Morganucodon*, Hopson itemizes five ways in which they are increasingly mammalian: (1) in the connection of the limbs, (2) in the mobility of the head, (3) in the fusing of the palate, (4) in the musculature of the jaw, and (5) in the migration of certain bones from the jaw to the middle ear. In response, Davis and Kenyon note parenthetically that soft tissues, such as those in the circulatory and reproductive systems, are not recorded in the fossils. Moreover, they argue, Hopson's series is not a lineage—a single path of genealogical descent—but merely a structural or morphological series. Quoting the biologist Douglas Futuyma, they note that it is impossible to tell which of the numerous therapsid species represented in the fossil record were in fact the ancestors of mammals, and consequently pose two questions. First, if only one therapsid lineage is ancestral to mammals, but several therapsid lineages have the same mammal-like features as the actual ancestral lineage, how powerful are those mammal-like features as evidence of ancestry? Second, if several therapsid lineages independently evolved into mammals, how plausible is it that they independently converged on the distinctive mammalian ear? To the suggestion that the mammalian ear might have been contained, unexpressed, in the genome of the earliest therapsid, they counter that natural selection acts only on expressed traits, and conclude that the evidence is in favor of the existence of "a common blueprint not developed by descent."

Readers are encouraged to read the source for themselves. It is Percival W. Davis and Dean H. Kenyon, *Of Pandas and People*, 2nd ed. (Dallas, TX: Haughton, 1993).

Gaps: The Evolutionary Biologist's View

Kevin Padian and Kenneth Angielczyk explain that modern paleontologists do not look for ancestors in the fossil record, but rather for transitional structures that illuminate the evolutionary history (phylogeny) of a lineage. They distinguish between lineal and collateral ancestors; the fossil record logically has far more of the former than of the latter, and much can be learned about a lineage's evolution from fossil species that are closely related, though not on the direct line of descent. Transitional fossils, then, become fossils that exhibit transitional features, rather than necessarily being the direct, lineal ancestors of later forms. In the article from which this excerpt is taken, the authors discuss a method of fossil analysis called *cladistics*, which groups organisms based on characteristics (anatomical, biochemical, embryological, etc.) that have been uniquely inherited from a common ancestor. It is a rigorous approach to hypothesizing and testing evolutionary relationships that has helped clarify the history of living things.

Are There Transitional Forms in the Fossil Record?

. . . Want to start a barroom fight? Ask another patron if he can produce proof of his unbroken patrilineal ancestry for the last four hundred years. Failing your challenge, the legitimacy of his birth is to be brought into question. At this insinuation, tables are overturned, convivial beverages spilled, and bottles fly. No fair, claims the gentle reader. This goes beyond illogic to impoliteness, because you are not only placing on the other patron an unreasonable burden of proof, you are questioning his integrity if he fails. But isn't that what creationists do when they claim that our picture of evolution in the fossil record must be fraudulent because we have so many gaps between forms? (p. 49).

. . . In the search for fossil forms that are ancestral to others, it is commonly assumed that such forms were the *actual individuals* from which living or later forms were descended. This definition is impossible to establish, unlikely on statistical grounds, and unnecessarily restrictive in concept (p. 54).

. . . Anthropologists distinguish . . . between *lineal* (direct) ancestors and *collateral* (side-branch) ancestors, and it is useful to borrow this concept to discuss real and apparent gaps in the fossil record. Collateral ancestors can still tell us much about the features, habits, and other characteristics of ancestors whose records may be lost but who would still be similar in most respects to those whose records we do have. Your grandfather is your lineal ancestor, whereas your great-uncle is a collateral ancestor; but were their lives and times necessarily much different? This is as true in paleontology as in anthropology. The most basal known member of a taxon (the one who retains the most "primitive" characteristics, and lacks the most derived ones) does not have to be the direct ancestor of the more derived ones; we can accept it as a collateral ancestor, and learn from it a great deal about the features of the actual (though hypothetical) unknown direct ancestor. However, we need to consider the most effective methods for approaching this kind of analysis (p. 55).

. . . Phylogenetic [cladistic] analysis, again, provides a solution. In the phylogenetic system, emphasis is placed not on discovering ancestral taxa, but on inferring ancestral (or general) and derived features. Shared derived features (synapomorphies) are the currency of phylogenetic reconstruction. If a synapomorphy is found in two or more related organisms, it is inferred to have been present in their common ancestor. (It could, of course, be independently evolved in each, and this question can be approached by adding more characters and taxa into the analysis.) So, rather than looking for fossils of lineal ancestors, we are now looking for synapomorphies that link collateral ancestors (p. 56).

. . . Among living terrestrial vertebrates there is perhaps no clade as distinctive and easily recognizable as the mammals. A variety of anatomical, physiological, osteological, and behavioral characteristics sets mammals apart from other groups of tetrapods. . . . The diagnosis given by Linnaeus when he coined the term Mammalia can still be used to differentiate between living mammals and other tetrapods. However, when we begin to take into account many of the early fossil relatives of mammals, things become much more confusing; it becomes harder to draw a clear distinction between what is a mammal and what is not. This problem stems in part from the fact that the various characteristics that so clearly delineate extant mammals did not all evolve at the same time. Instead, they evolved in a stepwise fashion, with some character states appearing before or after others (pp. 66–67).

. . . [Fossil relatives of mammals] are frequently referred to in the popular and scientific literature as "mammal-like reptiles," but this term is misleading and does not reflect our understanding of the relationships between mammals and reptiles. Mammals and the "mammal-like reptiles" are all members of the clade Synapsida and are characterized by having a single opening on the side of the skull through which jaw musculature passes. The clade Mammalia is hierarchically nested within Synapsida, and any synapsid that does not have the synapomorphies that diagnose mammals can be called a nonmammalian synapsid. Early nonmammalian synapsids are only somewhat similar to early reptiles, and not at all like extant ones. For example, their lower jaws are made up of a number of bones, like those of reptiles, instead of the single bone found in modern mammals. But these similarities were inherited by both lineages from their common amniote ancestor, so because they are shared primitive character states . . . they are not useful for grouping some synapsids within Reptilia and others within Mammalia. The lineal and collateral ancestors of mammals were never reptiles (reptiles are a separate lineage of amniotes), and the description of nonmammalian synapsids as "mammal-like reptiles" is a holdover of "ladder thinking." . . . Unfortunately, critics of evolution such as Denton and Johnson never bother to understand or clarify this distinction, because it is more to their purpose to suggest that we are vainly chasing non-existent transitions between "reptiles" and "mammals" than to show their audiences that the groups of mammals and their relative nestle nicely within a hierarchy of successively more inclusive phylogenetic groups (pp. 67–68).

Selection excerpted from:

Padian, Kevin, and Kenneth D. Angielczyk. 1999. Are There Transitional Forms in the Fossil Record? In *The Evolution-Creationism Controversy II: Perspectives on Science, Religion and Geological Education*, edited by P. H. Kelley, J. R. Bryan, and T. A. Hansen. Fayetteville, AR: Paleontological Society.

The following selection is excerpted from a much longer, more detailed, and extensively documented discussion of transitional fossils and their classification.

Common Descent, Transitional Forms, and the Fossil Record

Limits of the Fossil Record

. . . Soft-bodied or thin-shelled organisms have little or no chance of preservation, and the majority of species in living marine communities are soft-bodied. Consider that there are living today about 14 phyla of "worms" comprising nearly half of all animal phyla, yet only a few (e.g., annelids and priapulids) have even a rudimentary fossil record.

. . . Even those organisms with preservable hard parts are unlikely to be preserved under "normal" conditions. Studies of the fate of clamshells in shallow coastal waters reveal that shells are rapidly destroyed by scavenging, boring, chemical dissolution, and breakage. Occasional burial during major storm events is one process that favors the incorporation of shells into the sedimentary record, and their ultimate preservation as fossils. Getting terrestrial vertebrate material into the fossil record is even more difficult. The terrestrial environment is a very destructive one: with decomposition and scavenging together with physical and chemical destruction by weathering.

The limitations of the vertebrate fossil record can be easily illustrated. The famous fossil *Archaeopteryx*, occurring in a rock unit renowned for its fossil preservation, is represented by only seven known specimens, of which only two are essentially complete. Considering how many individuals of this genus probably lived and died over the thousands or millions of years of its existence, these few known specimens give some feeling for how few individuals are actually preserved as fossils and subsequently discovered. . . . Complete skeletons are exceptionally rare. For many fossil taxa, particularly small mammals, the only fossils are teeth and jaw fragments. If so many fossil vertebrate species are represented by single specimens, the number of completely unknown species must be greater still!

. . . In addition to these preservational biases, the erosion, deformation, and metamorphism of originally fossiliferous sedimentary rocks has eliminated significant portions of the fossil record over geologic time. Furthermore, much of the fossil-bearing sedimentary record is hidden in the subsurface, or located in poorly accessible or little studied geographic areas. For these reasons, only a small portion of those once living species actually preserved in the fossil record have been discovered and described by science (pp. 154–156).

. . . Climbing Down the Tree of Life

. . . A long-standing misperception of the fossil record of evolution is that fossil species form single lines of descent with unidirectional trends. Such a simple linear view of evolution is called orthogenesis ("straight origin"), and has been rejected by paleontologists as a model of evolutionary change. The reality is much more complex than that, with numerous branching lines of descent and multiple anatomical trends. The fossil record reveals that the history of life can be understood as a densely branching bush with many short branches (short-lived lineages). The well-known fossil horse series, for example, does not represent a single, continuous, evolving lineage. Rather, it records more or less isolated twigs of an adapting and diversifying limb of the tree of life. While incomplete, this record provides important insights into the patterns of morphological divergence and the modes of evolutionary change.

Curiously, some critics of evolution view the record of fossil horses from *Hyracotherium* ("Eohippus"), the earliest known representative of this group, to the modern *Equus* as trivial. However, that is only because the intermediate forms are known. Without them, the anatomical gap would be very great. *Hyracotherium* was a very small (some species only 18 inches long) and generalized herbivore (probably a browser). In addition to the well-known difference in toe number (4 toes in front, 3 in back), *Hyracotherium* had a narrow, elongate skull with a relatively small brain and eyes placed well forward in the skull. It possessed small canine teeth, simple tricuspid premolars, and low-crowned simple molars. Over geologic time and within several lines of descent, the skull became much deeper, the eyes moved back, and the brain became larger. The incisors were widened, premolars took the form of molars, and both premolars and molars became very high-crowned with a highly complex folding of the enamel.

The significance of the fossil record of horses becomes clearer when it is compared with that of the other members of the odd-toed ungulates (hoofed mammals). The fossil record of the extinct brontotheres is quite good, and the earliest representatives of this group are very similar to *Hyracotherium*. Likewise, the earliest members of the tapirs and rhinos were also very much like the earliest horses. All these very distinct groups of terrestrial vertebrates can be traced back through a sequence of forms to a group of very similar small, generalized ungulates in the early Eocene. The fossil record thus supports the derivation of horses, rhinos, tapirs and brontotheres from a common ancestor resembling *Hyracotherium*. Furthermore, moving farther back in time to the late Paleocene, the earliest representatives of the odd-toed ungulates, even-toed ungulates (deer, antelope, cattle, pigs, sheep, camels, etc.), and the proboscideans (elephants and their relatives) were also very similar to each other.

Similar patterns are seen when looking at the fossil record of the carnivores. One group of particular interest is the pinnipeds (seals, sea lions, and walruses). These aquatic carnivores have been found to be closely related to the bears, and transitional forms are known from the early and middle Miocene. More broadly, the living groups of carnivores are divided into two main branches, the Feliformia (cats, hyenas, civets, and mongooses) and the Caniformia (dogs, raccoons, bears, pinnipeds, and weasels). The earliest representatives of these two carnivore branches are very similar to each other, and likely derived from a primitive Eocene group called the miacids. Of the early carnivores, an eminent vertebrate paleontologist has stated: "Were we living at the beginning of the Oligocene, we should probably consider all these small carnivores as members of a single family." This statement also illustrates the point that the erection of a higher taxon is done in retrospect, after sufficient divergence has occurred to give particular traits significance (pp. 162–166).

. . . The complex of transitional fossil forms has created significant problems for the definition of the class Mammalia. For most workers, the establishment of a dentary-squamosal jaw articulation is considered one of the primary defining characters for mammals. The transition in jaw articulation associated with the origin of mammals is particularly illustrative of the appearance of a "class-level" morphologic character. In nonmammalian vertebrates, the lower jaw contains several bones, and a small bone at the back of the jaw (the articular) articulates with a bone of the skull (the quadrate). In mammals, the lower jaw consists of only a single bone, the dentary, and it articulates with the squamosal bone of the skull. Within the cynodont lineage, the dentary bone becomes progressively larger and the other bones are reduced to nubs at the back. In one group of advanced cynodonts, the dentary bone has been brought nearly into contact with the squamosal. The earliest known mammals, the morganucodonts, retain the vestigial lower jaw bones of the earlier cynodonts. These small bones still formed a reduced, but functional, jaw joint adjacent to the new dentary-squamosal mammalian articulation. These animals possessed simultaneously both "reptilian" and mammalian jaw articulations! The "reptilian" jaw elements were subsequently detached completely from the jaw to become the bones of the mammalian middle ear. Better intermediate character states could hardly be imagined!

As with most transitions between higher taxonomic categories, there is more than one line of descent that possesses intermediate morphologies. Again, this is consistent with both the expectations of evolutionary theory and the nature of the fossil record. The prediction would be for a bush of many lineages, most of which would be dead ends (p. 168).

[References and illustrations omitted; see original article.]

Selection excerpted from:

Miller, Keith B. 2003. Common Descent, Transitional Forms, and the Fossil Record. In *Perspectives on an Evolving Creation*, edited by K. B. Miller. Grand Rapids, MI: Eerdmans.

• •

THE CAMBRIAN EXPLOSION

The Cambrian Explosion began about 535 million years ago when basic features of "body plans" of invertebrates first appear in the fossil record—shells as found in mollusks, jointed limbs as found in arthropods, exoskeletons, and so on. Most Precambrian animal fossils looked quite different from living invertebrates, although many relatives of modern forms occur before the Cambrian. Creationists believe that the rapidity with which the Cambrian fauna appear rules out the possibility of natural selection producing these varieties; to them, God created the different "kinds" separately. Paleontologists consider the Cambrian Explosion to be an interesting scientific puzzle, but by no means a "problem" for evolution or even evolution by natural selection. Most now conclude that invertebrate body plans have a history extending well before the Precambrian/Cambrian boundary—though fossil evidence for this is scarce. But as will be noted below, the fossil record is not the only source of information on relationships among invertebrate groups.

The Cambrian as Viewed by Creationists

The author of the next selection is the foremost debater promoting Young Earth Creationism. (Intelligent Design proponents hold similar views.)

Attack and Counterattack: The Fossil Record

. . . There are two huge gaps in the fossil record that are so immense and indisputable that any further discussion of the fossil record becomes superfluous. These are the gap between microscopic, single-celled organisms and the complex, multicellular invertebrates, and the vast gap between these invertebrates and fish. There are now many reports in the scientific literature claiming the discovery of fossil bacteria and algae in rocks supposedly as old as 3.8 billion years. Paleontologists generally consider that the validity of these claims is beyond dispute. In rocks of the so-called Cambrian Period, which evolutionists believe began to form about 600 million years ago, and which supposedly formed during about 80 million years, are found the fossils of a vast array of very complicated invertebrates—sponges, snails, clams, brachiopods, jellyfish, trilobites, worms, sea urchins, sea cucumbers, sea lilies, etc. Unnumbered billions of these fossils are known to exist. Supposedly, these complex invertebrates had evolved from a single-celled organism.

The rocks that generally underlie the Cambrian rocks are simply called Precambrian rocks. Some are thousands of feet thick, and many are undisturbed—perfectly suitable for the preservation of fossils. If it is possible to find fossils of microscopic, single-celled, soft-bodied bacteria and algae, it should certainly be possible to find fossils of the transitional forms between those organisms and the complex invertebrates. Many billions times billions of the intermediates would have lived and died during the vast stretch of time required for the evolution of such a diversity of complex organisms. The world's museums should be bursting at the seams with enormous collections of the fossils of transitional forms. As a matter of fact, not a single such fossil has ever been found! Right from the start, jellyfish have been jellyfish, trilobites have been trilobites, sponges have been sponges, and snails have been snails. Furthermore, not a single fossil has been found linking, say, clams and snails, sponges and jellyfish, or trilobites and crabs, yet all of the Cambrian animals supposedly have been derived from common ancestors.

For a time, evolutionists believed that the Ediacaran Fauna, originally discovered in Australia but now known to be worldwide in distribution, contained creatures that, even though already very complex in nature, might be ancestral to many of the Cambrian animals. Some of the Ediacaran creatures were placed in the same categories as the Cambrian jellyfish, worms, and corals. According to Adolph Seilacher, a German paleontologist, the Ediacaran creatures are, however, basically different from all of the Cambrian animals, and so could not possibly have been ancestral to them. It is believed that all of the Ediacaran creatures became extinct without leaving any evolutionary offspring (Gould 1984). Thus, the Cambrian "explosion," as it is commonly called, remains an unsolved mystery for evolutionists (pp. 115–116).

. . . Eldredge's main argument is that evolution does not necessarily proceed slowly and gradually, but that some episodes in evolution may, geologically speaking, proceed very rapidly (Eldredge 1982). Thus, just before the advent of the Cambrian, for some reason or other, there was an evolutionary burst—a great variety of complex multicellular organisms, many with hard parts, suddenly evolved. This evolution occurred so rapidly (perhaps in a mere fifteen to twenty million years, more or less) there just wasn't enough time for the intermediate creatures to leave a detectable fossil record.

This notion of explosive evolution is really not a new idea at all, as it has been employed in the past to explain the absence of transitional forms (Simpson 1949). This notion will not stand up under scrutiny, however. First, what is the only evidence for these postulated rapid bursts of evolution? The absence of transitional forms! Thus, evolutionists, like Eldredge, Simpson, and others, are attempting to snatch away from creation scientists what these scientists consider to be one of the best evidences for creation, that is, the absence of transitional forms, and use it as support for an evolutionary scenario! (pp. 117–118).

. . . Later in the book by Eldredge quoted above, Eldredge suggests the most incredible notion of all to explain away the vast Cambrian explosion. He states:

> We don't see much evidence of intermediates in the Early Cambrian because the intermediates had to have been soft-bodied, and thus extremely unlikely to become fossilized. (Eldredge 1982: 130)

It is difficult to believe that Eldredge or any other scientist could have made such a statement. Whatever they were, the evolutionary predecessors of the Cambrian animals had to be complex. A single-celled organism could not possibly have suddenly evolved into a great variety of complex invertebrates without passing through a long series of intermediates of increasing complexity. Surely, if paleontologists are able to find numerous fossils of microscopic, single-celled, soft-bodied bacteria and algae, as Eldredge does not doubt they have, then they could easily find fossils of all the stages intermediate between these microscopic organisms and the complex invertebrates of the Cambrian. Furthermore, in addition to the many reported findings of fossil bacteria and algae, there must be many hundreds of finds of soft-bodied, multicellular creatures, such as worms and jellyfish, in the scientific literature. The creatures of the Ediacaran Fauna, which have been reported from five continents, are soft-bodied (pp. 118–119).

REFERENCES CITED

Eldredge, Niles. 1982. *The Monkey Business*. New York: Washington Square Press.
Gould, S. J. 1984. The Ediacaran experiment: Insights into mass extinction theory. *Natural History* 93: 14.
Simpson, G. G. 1949. *The Meaning of Evolution*. New Haven, CT: Yale University Press.

Selection excerpted from:

Gish, Duane T. 1993. *Creation Scientists Answer Their Critics*. El Cajon, CA: Institute for Creation Research.

The Cambrian as Viewed by Evolutionary Biologists

Jonathan Wells's book Icons of Evolution *(2002) discusses the Cambrian Explosion as a "problem for Darwinian evolution," and raises many of the same points made in Duane Gish's selection above. Paleontologist Alan Gishlick's response to Wells's treatment of the Cambrian addresses several creationist claims, arguing that antievolutionists misstate or misunderstand the scientific evidence of the fossil record and the rules of taxonomy. Evolutionary biologists consider the Cambrian Explosion an interesting scientific puzzle to be solved, not "evidence against evolution."*

The Cambrian Explosion

. . . The gist of Wells's argument is that the Cambrian Explosion happened too fast to allow large-scale morphological evolution to occur by natural selection ("Darwinism"), and that the Cambrian Explosion shows "top-down" origination of taxa ("major" "phyla" level differences appear early in the fossil record rather than develop gradually), which he claims is the opposite of what evolution predicts. He asserts that phylogenetic trees predict a different pattern for evolution than what we see in the Cambrian Explosion. These arguments are spurious and show his lack of understanding of basic aspects of both paleontology and evolution.

Wells mistakenly presents the Cambrian Explosion as if it were a single event. The Cambrian Explosion is, rather, the preservation of a series of faunas that occurs over a 15–20 million year period starting around 535 million years ago (MA). A fauna is a group of organisms that live together and interact as an ecosystem; in paleontology, "fauna" refers to a group of organisms that are fossilized together because they lived together. The first fauna that shows extensive body plan diversity is the Sirius Passet fauna of Greenland, which is dated at around 535 MA. The organisms preserved become more diverse by around 530 MA, as the Chenjiang fauna of China illustrates. . . . The diversification continues through the Burgess shale fauna of Canada at around 520 MA, when the Cambrian faunas are at their peak. Wells makes an even more important paleontological error when he does not explain that the "explosion" of the middle Cambrian is preceded by the less diverse "small shelly" metazoan faunas, which appear at the beginning of the Cambrian (545 MA). These faunas are dated to the early Cambrian, not the Precambrian as stated by Wells. This enables Wells to omit the steady rise in fossil diversity between the beginning of the Cambrian and the Cambrian Explosion.

In his attempt to make the Cambrian Explosion seem instantaneous, Wells also grossly mischaracterizes the Precambrian fossil record. In order to argue that there was not enough time for the necessary evolution to occur, Wells implies that there are no fossils in the Precambrian record that suggest the coming diversity or provide evidence of more primitive multicellular animals than those seen in the Cambrian Explosion. . . . Wells . . . asserts that there is no evidence for metazoan life until "just before" the Cambrian Explosion, thereby denying the necessary time for evolution to occur. Yet Wells is evasive about what counts as "just before" the Cambrian. Cnidarian and possible arthropod embryos are present 30 million years "just before" the Cambrian. There is also a mollusc, Kimberella, from the White Sea of Russia dated approximately 555 million years ago, or 10 million years "just before" the Cambrian. This primitive animal has an uncalcified "shell," a muscular foot and a radula inferred from "mat-scratching" feeding patterns surrounding fossilized individuals. These features enable us to recognize it as a primitive relative of molluscs, even though it lacks a calcified shell. There are also Precambrian sponges as well as numerous trace fossils indicating burrowing by wormlike metazoans beneath the surface of the ocean's floor. Trace fossils demonstrate the presence of at least one ancestral lineage of bilateral animals nearly 60 million years "just" before the Cambrian. Sixty million years is approximately the same amount of time that has elapsed since the extinction of non-avian dinosaurs, providing plenty of time for evolution. In treating the Cambrian Explosion as a single event preceded by nothing, Wells misrepresents fact—the Cambrian Explosion is not a single event, nor is it instantaneous and lacking in any precursors.

. . . Wells invokes a semantic sleight of hand in resurrecting a "top-down" explanation for the diversity of the Cambrian faunas, implying that phyla appear first in the fossil record, before lower categories. However, his argument is an artifact of taxonomic practice, not real morphology. In traditional taxonomy, the recognition of a species implies a phylum. This is due to the rules of taxonomy, which state that if you find a new organism, you have to assign it to all the necessary taxonomic ranks. Thus when a new organism is found, either it has to be placed into an existing phylum or a new one has to be erected for it. Cambrian organisms are either assigned to existing "phyla" or new ones are erected for them, thereby creating the effect of a "top-down" emergence of taxa.

. . . [T]he "higher" taxonomic groups appear at the Cambrian Explosion . . . because the Cambrian Explosion organisms are often the first to show features that allow us to relate them to living groups. The Cambrian Explosion, for example, is the first time we are able to distinguish a chordate from an arthropod. This does not mean that the chordate or arthropod lineages evolved then, only that they then became recognizable as such.

. . . Similarly, before the Cambrian Explosion, there were lots of "worms," now preserved as trace fossils (i.e., there is evidence of burrowing in the sediments). However, we cannot distinguish the chordate "worms" from the mollusc "worms" from the arthropod "worms" from the worm "worms." Evolution predicts that the ancestor of all these groups was wormlike, but which worm evolved the notochord, and which the jointed appendages? . . . If the animal does not have the typical diagnostic features of a known phyla [*sic*], then we would be unable to place it and (by the rules of taxonomy) we would probably have to erect a new phylum for it. When paleontologists talk about the "sudden" origin of major animal "body plans," what is "sudden" is not the appearance of animals with a particular body plan, but the appearance of animals that we can recognize as having a particular body plan. Overall, however, the fossil record fits the pattern of evolution: we see evidence for wormlike bodies first, followed by variations on the worm theme. Wells seems to ignore a growing body of literature showing that there are indeed organisms of intermediate morphology present in the Cambrian record and that the classic "phyla" distinctions are becoming blurred by fossil evidence.

Finally, the "top-down" appearance of body plans is, contrary to Wells, compatible with the predictions of evolution. The issue to be considered is the practical one that "large-scale" body-plan change would of course evolve before minor ones. (How can you vary the lengths of the beaks before you have a head?) The difference is that many of the "major changes" in the Cambrian were initially minor ones. Through time they became highly significant and the basis for "body plans." For example, the most primitive living chordate Amphioxus is very similar to the Cambrian fossil chordate Pikaia. Both are basically worms with a stiff rod (the notochord) in them. The amount of change between a worm and a worm with a stiff rod is relatively small, but the presence of a notochord is a major "body-plan" distinction of a chordate. Further, it is just another small step from a worm with a stiff rod to a worm with a stiff rod and a head (e.g., Haikouella) or a worm with a segmented stiff rod (vertebrae), a head, and fin folds (e.g., Haikouichthyes). Finally add a fusiform body, fin differentiation, and scales: the result is something resembling a "fish" (Figure 8.1). But, as soon as the stiff rod evolved, the animal was suddenly no longer just a worm but a chordate—representative of a whole new phylum! Thus these "major" changes are really minor in the beginning, which is the Precambrian–Cambrian period with which we are concerned.

[Readers are encouraged to check the original article for extensive references, omitted here.]

Selection excerpted from:

Gishlick, Alan D. 2003. *Icons of Evolution? Why Much of What Jonathan Wells Writes About Evolution Is Wrong.* Icon #2: Darwin's "Tree of Life." National Center for Science Education, Inc., 2003. Accessed September 1, 2003. Available from http://www.ncseweb.org/icons/icon2tol.html.

The fossil record is not the only source of insight into the Cambrian Explosion. Scientists also have evidence from developmental biology—the study of how a fertilized egg turns into an organism. Some early-acting genes can have profound effects on a developing embryo, and changes in these genes are inferred to have initiated the body plan changes of the Cambrian. Working together, molecular biologists and paleontologists are starting to untangle a fascinating scientific puzzle.

Figure 8.1
Evolutionary Changes. Courtesy of Alan Gishlick

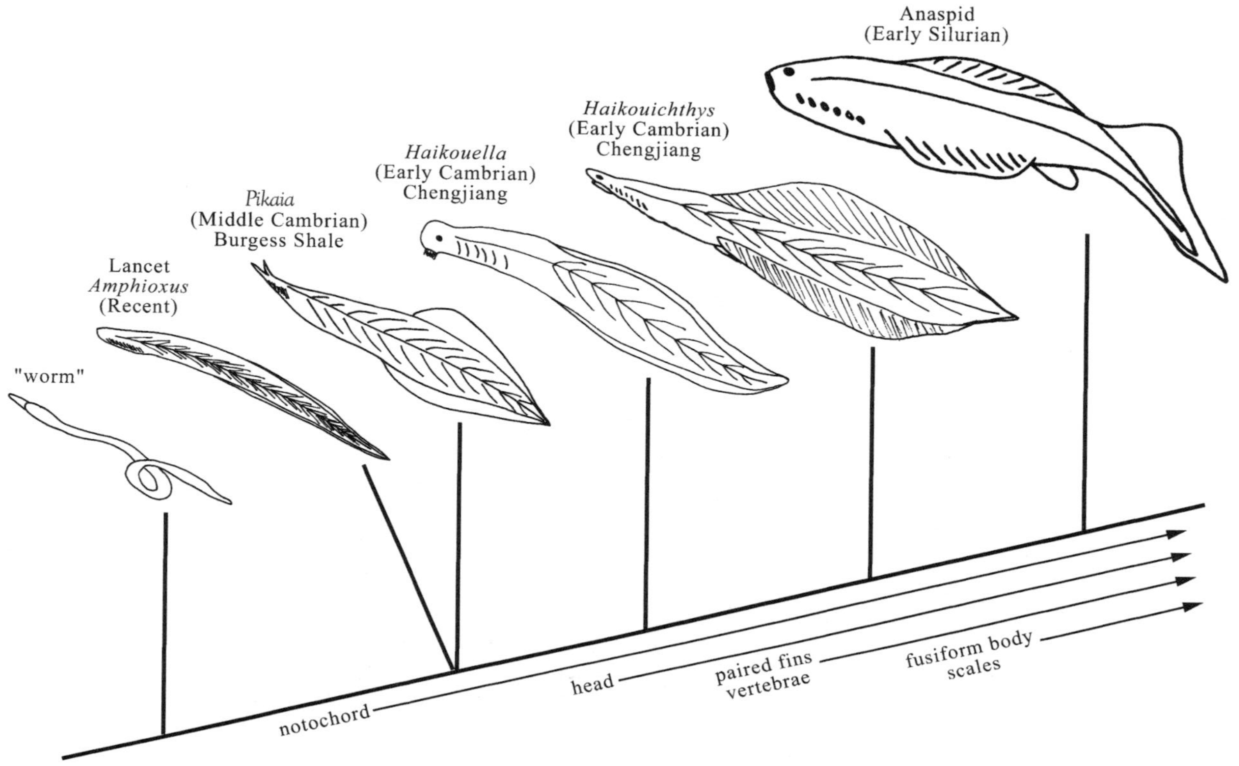

When We Were Worms

. . . The paleontologists struggling to explain the Cambrian explosion face a tough task—500 million years separate them from their subject. Until recently their only option was to study the animal fossil record. But now, more and more researchers are taking a different path to enlightenment: scrutinizing the genetic record that has been handed down through the ages to today's creatures. Comparing the genes of living animals has enabled biologists to crawl back down the evolutionary tree to deduce which genes were present back then and what roles they might have played.

The exploration has led some researchers to make the heretical claim that the Cambrian explosion never happened, and others to say it certainly did, and that they have found the likely mechanism: the genes that help the cells of a developing embryo know front from back, top from bottom, and near from far. They believe that these genes were a necessary prerequisite for the explosion—though they may not have been what set it off.

The most famous of these genes belongs to a set known as the hox cluster. Hox genes are the mapmakers that tell the embryo's cells where they are on the body's front-to-back axis and thus what they should become. In fruit flies, in which they were first discovered, eight hox genes line up on their chromosome like a train of boxcars. The gene at the front of the train tells cells there to make a head and other paraphernalia characteristic of that part of the body, number two takes over a bit further back, and so on back to the guard's van (Americans call it the caboose), which holds sway over the hindmost end of the animal. By mucking up this orderly sequence, researchers create startling freaks such as flies sprouting legs from their heads. Other sets of genes lay out the body's up-down axis or distinguish the base of a leg or a wing from its tip.

. . . And about two years ago, three paleontologists began to suspect that these genes could be the answer to another great enigma, the Cambrian explosion. The modular way in which the genes map body regions would have provided evolution with a mechanism to modify one part of the body without changing the rest, to duplicate segments, or to add new appendages where none existed before. This is the line taken by James Valentine, of the University of California at Berkeley, David Jablonski of the University of Chicago, and Douglas Erwin of the National Museum of Natural History in Washington, D.C.

. . . Valentine, Jablonski, and Erwin needed to show that these mapmaking genes actually existed in the Cambrian. That posed a problem—Jurassic Park notwithstanding, genes don't fossilise, least of all for half a billion years and more.

. . . Fortunately, living organisms hold much of the secret. "The present is the key to the past," says Rudolf Raff, an evolutionary developmental biologist at Indiana University in Bloomington. Raff's reasoning is simple. Living organisms are the tips of branches in the evolutionary tree. If the same gene occurs in two of these animals, that gene must also have been present in their common ancestor, back where their lineages first split apart. By comparing ever more distantly related organisms, biologists can move down toward the earliest branches near the tree's root.

Fruit flies and frogs—two veterans of the biology labs—have very different body designs. And they occupy twigs on two of the most fundamental branches of the animal tree, the groups known to biologists as protostomes and deuterostomes, respectively (named after the way in which the embryonic mouth forms). Paleontologists know from the fossil record that these two lineages diverged no later than the early Cambrian, 535 million years ago. They could scarcely be more distant cousins.

. . . Yet over the past few years, developmental biologists have accumulated more and more evidence of astounding similarities between the DNA sequences in the mapmaking genes of these two groups. Flies and frogs, and by inference their common ancestor in the Cambrian or before, share six hox genes. . . . [T]he two have diverged so little in more than half a billion years that scientists can snip the gene out of a fly, plug it into a frog, and it will work perfectly, triggering the development of the bottom half of a frog wherever you insert it in the embryo. . . . Even genes that dictate the development of such modern-seeming accoutrements as hearts, nervous systems, and body segments appear to have been present in the frog-fruit fly common ancestor, way back in the Cambrian or earlier.

This was just the evidence the three paleontologists needed—and its implications are still sinking in. For a start, it questions the old assumption that the common ancestor of horses and horseflies, lobsters and Londoners—the giga-great-grandmother of the early Cambrian or before—was a "roundish flatworm," little more than an oozing blob of cells. After all, if the genetic evidence is to be believed, that ancestor could equally well have been a sophisticated, segmented worm with eyes, a heart, a nervous system, possibly even antennae or legs, says Eddy De Robertis, an embryologist at the University of California at Los Angeles.

Another explanation, and the one that fits in best with Valentine and Co's hypothesis, is that those key developmental genes existed in that roundish flatworm, but didn't trigger the growth of eyes, limbs, hearts, and segments as they do in post-Cambrian animals. . . . The Precambrian "eye" that pax6 helped form may have been nothing more than a crude photosensor with a pigment cell to back it up, for instance. Indeed, the fossil record shows no trace of anything as sophisticated as the insect and vertebrate eyes in the earliest protostomes and deuterostomes.

. . . John Finnerty and Mark Martindale of the University of Chicago report that sea anemones have a rich set of hox genes, even though these most primitive of animals have no head nor tail—they face the world equally well in all directions. . . . Finnerty suspects that [hox genes] may map out the anemones' far simpler up-and-down axis instead. The best guess is that the Precambrian flatworm then commandeered these crude mapmaking genes for use in the crucial front-to-back axis of its more complex body plan—possibly their first big role as the "language of evolution."

"Creationists have often said that the one thing we can't explain is the extremely rapid appearance of body plans in the Cambrian," says Valentine. "And this is the answer. We haven't understood it until these development guys."

. . . [W]hatever happened to kick-start the Precambrian worm into an evolutionary frenzy, it probably found it much easier to spin off new body plans than a creature would today. "You can do a lot to that animal in terms of its basic body arrangement for two reasons," says Raff. "One is that the world is empty ecologically, so there can be a lot of experiments. Whatever you're making doesn't have to be very good, because there isn't much competition. The second thing is that because the body's relatively simple, changes that would now be horrendous—like changing dorsal and ventral—would have been nothing at all.

"In the post-Cambrian world, competition is severe, so if you're not good at making a living, you're dead meat. And body plans have become more elaborate, so changes that you make have more consequences. You've connected the genetic machinery together in a certain way, and it may be hard to unconnect it. That's left us with a world in which evolution is largely within body plans."

In short, animal life on Earth has grown up. The heady, experimental days of its carefree youth are past and now, saddled with a job, a mortgage, responsibilities, it has settled down into a steady, plodding respectability. Looking back now, we can shed a tear for those days so long ago when life was young and we were worms.

Selection excerpted from:

Holmes, Bob. 1997. When We Were Worms. *New Scientist* 156 (2104): 30–35.

• •

DESIGN

Living things often show remarkable fit to environments: the grasping foot of a wren allows it to perch on a branch, while the webbed foot of a duck allows it efficiently to propel itself through water. As discussed in chapter 4, before Darwin, this fit of organisms to their ways of life was explained through special creation by God; one of the early religious objections to natural selection was that it replaced the necessity of the direct hand of God for design. Many antievolutionists continue to believe that if natural selection *could* explain this sort of design, then God is necessarily removed from creation. So the Argument from Design remains an important component of the creation and evolution controversy.

Creationists regularly (and incorrectly) equate evolution with chance, and since it is absurd to imagine that the complexity of the universe and living things could have come about though random behavior of matter, evolution is rejected as being too improbable. They believe that it is more probable that adaptation and complexity are the result of the purposeful design of a Creator. Both traditional creationists and (especially) Intelligent Design creationists argue that the origin of life, and anatomical or biochemical/cellular structural complexity, are so complex that natural processes cannot explain them.

Arguments Against Evolution Producing Design

Oller presents a common improbability argument raised by YECs.

Not According to Hoyle

. . . In his well-illustrated and impressive new book, *The Intelligent Universe* (London: Michael Joseph, 1983, 256 pp.), Hoyle says:

> As biochemists discover more and more about the awesome complexity of life, it is apparent that its chances of originating by accident are so minute that they can be completely ruled out. Life cannot have arisen by chance. (pp. 11–12)

Does this mean that Hoyle has become a creationist? Well, not exactly, and he doesn't expect to either. To forestall any speculation about his apparent "conversion," he says bluntly: "I am not a Christian, nor am I likely to become one as far as I can tell" (p. 251). Still, Hoyle argues that there must have been some "intelligence" behind the emergence of life on Earth. Setting aside the question of what sort of intelligence, he offers an interesting line of argument.

The probability that the simplest life-form could just accidentally arrange itself from particles floating in an ideally prepared primordial soup is very slim. To appreciate just how slim, Hoyle proposes an analogy. He asks how long it would take a blindfolded person to solve a Rubik Cube. Suppose he worked very fast; say, a move a second without resting. According to Hoyle's figuring it would take approximately 67.5 times the estimated age of the universe (allowing the generous figure of 20 billion years since the big bang) for him to reach a solution—about 1.35 trillion years. Judging from the life expectancy of human beings we could say that a solution of the Rubik Cube could not be achieved at all by a blindfolded person. Yet this is just about the same difficulty as the accidental formation of just one of the chains of amino acids necessary to living cells. In the human cell, Hoyle points out, there are about 200,000 such proteins. The chance of getting all 200,000 by accident is really small. In fact, even if an ideal primordial soup existed, and if it were repeatedly jolted by electrical charges (as in the famous Miller-Urey experiment), the time required for the formation of any one of the requisite 200,000 proteins would be roughly equivalent to 293.5 times the estimated age of the Earth (set at the standard 4.6 billion years).

Yet the odds against the accidental formation of a living organism are considerably worse than the odds against a blindfolded solution of the Rubik Cube—the latter being estimated by Hoyle to be about 50 billion trillion to 1. The trouble is that even a simple protozoan, or a bacterium, requires the prior formation of about 2,000 enzymes, themselves also complex proteins, which are critical to the successful formation of all the other 198,000 or so requisite proteins. The odds in favor of the accidental formation of all 2,000 by accident (never mind the other 198,000), without which no living organism could have come into existence, approaches a truly infinitesimal magnitude. The odds would be similar to those against 2,000 blindfolded persons working Rubik Cubes independently and just accidentally coming to perfect solutions simultaneously—according to Hoyle, roughly 1,040,000 to 1. Or, to give a more graspable notion of the improbability, Hoyle says, it would be roughly comparable to rolling double-sixes 50,000 times in a row with unloaded dice. Looking at it from the point of view of the expected time lapse before reaching a solution, the predicted heat death of our solar system would have occurred early on, and our Milky Way galaxy would have rolled itself up like a scroll long before a solution could be hoped for. . . .

[References omitted.]

Selection excerpted from:

Oller, John W., Jr. 1984. Not According to Hoyle. *Impact* 138: 1–4.

Michael Behe is a leading proponent of the concept of irreducible complexity, discussed in chapter 6. Our intent was to reprint "The Challenge of Irreducible Complexity" (*Natural History*, April 2002, p. 74), but Dr. Behe denied permission for this use.

In this article, Behe begins by explaining the term "black box" as "a system whose inner workings are unknown"; in Darwin's day, he says, the cell was effectively a black box. Today, however, it is known that the cell contains structures of astonishing complexity; Behe poses the question of how to ascertain whether Darwinian natural selection is capable of accounting for it. Quoting Darwin's acknowledgment "If it could be demonstrated that any complex organ existed which could not possibly have been formed by numerous, successive, slight modifications, my theory would absolutely break down," Behe introduces the concept of irreducible complexity: a system is irreducibly complex if it would not be able to perform its function if any of its parts were removed. His example is the mousetrap. Behe argues that it is extremely unlikely for such systems to be produced via Darwinian natural selection because any precursor that lacked a part would have been unable to perform its function and therefore would have probably been selected against. There are irreducibly complex systems at the cellular level, however, such as the bacterial flagellum (the whiplike propeller possessed by certain bacteria), the system by which proteins flow throughout eukaryotic cells, and the blood-clotting mechanism. Although these systems are described in biochemistry textbooks and journals, there is very little information about their supposed evolution by natural selection. Behe ends by expressing his optimism that eventually the hypothesis that they are the products of intelligent design will be seriously entertained by the scientific community.

Readers are encouraged to read the source firsthand and also Michael Behe, *Darwin's Black Box: The Biochemical Challenge to Evolution* (New York: Free Press, 1996). More detail on Behe's concept of irreducible complexity can be found in the same volume.

Rebuttals to the Design Argument

Biologist Kenneth R. Miller has debated creationists of both the Young Earth and Intelligent Design persuasions. In the following selection he comments on a debate he had with the Institute for Creation Research's star debater, Duane Gish.

The Laws of Probability

The next argument was an old standard. Dr. Gish noted the great complexity of living cells and the various other forms of life on earth. He argued that the mathematics of probability would render it impossible for life to develop from nonlife all by itself, no matter how much time was allowed:

> Most proteins consist of several hundred amino acids, each arranged in precise sequence, and DNA and RNA usually consist of thousands of nucleotides also arranged in precise order. The number of different possible ways these subunits can be arranged is so incredibly astronomical that it is literally impossible for a single molecule of protein or DNA to have been generated by chance in five billion years.

He backed up this claim by citing calculations by Hubert Yockey. But these calculations are based on two false assumptions which stack the deck against evolution: first, that a particular nucleotide or amino acid sequence must assemble completely by chance—and only that specific sequence will be accepted—and, second, that no small nucleotide chains are capable of self-replication.

Yet, in the globin protein sequence (the polypeptide part of hemoglobin) only seven amino acids, out of more than one hundred, are always the same when we examine the many globins which are used by different organisms. If the creationist calculations are done with this fact in mind, we would discover that such sequences form very quickly. Second, the sequences would not have to assemble from scratch. Recent work by Orgel and Eigen and others has shown that RNA nucleotides can spontaneously form small chains. Furthermore, these small chains can proceed to self-replicate. Often when such organic molecules get to be twenty to twenty-five amino acids long, they can spontaneously double their lengths through this replication process. (Indeed, many of the molecules found in living things bear evidence of having evolved in exactly this way.) The net result is thousands and thousands of variant copies being produced quickly. Therefore, the sequences that Dr. Gish says could never form would in fact self-assemble in a few months or years, given the whole earth as a laboratory. Since Yockey's calculations do not allow for this replication, his mathematical results are light years away from the truth.

Gish argued next that hundreds of different functional proteins would have had to form simultaneously. He assumed that this also would be another impossibility. Yet, there are numerous papers with copious data showing that the many modern proteins appear to have derived from a few ancestral proteins. He also assumed that, if modern cells have two hundred proteins, the earliest protocells also had two hundred proteins. A wealth of experimental results refutes that assumption as well. However, in spite of the open availability of all this data, the creationists go right on making these same tired old statements. . . .

Selection excerpted from:

Miller, Kenneth R. 1982. Answers to the Standard Creationist Arguments. *Creation/Evolution* 3 (1): 7–8.

Douglas Futuyma answers design arguments of the sort that Behe presented above. His answer is part of a larger series of criticisms of Darwin's Black Box, *which includes replies by Behe and other design proponents. The full discussion is online at http://bostonreview.net/BR21.6/orr.html.*

Miracles and Molecules

. . . Complex biochemical systems, then, bear the molecular stamp of their evolutionary origins. Often, these systems can be found, in one or another organism, in a primitive, less complex state—a state that functions adequately, even if not as efficiently as the more complex state that evolved in other lineages. The eye of a mammal is wondrously, perhaps "irreducibly," complex, but an eye without a lens, capable at least of distinguishing light from dark, is better than no eye at all. Likewise, a lamprey's hemoglobin, even if less efficient than that of a jawed vertebrate, suffices to keep lampreys alive. Yet it is doubtful that a mammal could survive with a lamprey-like hemoglobin, for the physiological functions that have evolved in mammals, such as maintaining high body temperature, demand oxygen at a rate that can be supplied only by more efficient, tetrameric [four-strand] hemoglobin. Likewise, it is unlikely that a mammalian fetus could survive without its special hemoglobin. What was once merely an advantage has become a necessity. As Orr emphasizes, irreducible complexity is acquired—it evolves.

Among vertebrates, only a subset—the jawed vertebrates that first evolved about 430 million years ago—have tetrameric hemoglobin, and of these only a subset—the mammals whose ancestors became differentiated from other reptiles about 320 million years ago—have fetal hemoglobin. These facts permit two possible explanations. One—Behe's explanation—is that the common ancestor of all vertebrates, or of all life, was equipped with all the molecular machinery any of its descendants would ever use, and that most of the machinery was lost in most lineages. This hypothesis is not only ludicrous, but also, as Orr points out, makes predictions that are contradicted by evidence. The alternative hypothesis is that new molecular complexities came into existence in various lineages of organisms at different points in time.

If this is true, and if we were to follow Behe in denying a natural, evolutionary origin of each such instance, then each origin of a divergent, duplicate hemoglobin requires us to postulate a special intervention by the omnipotent designer. Bear in mind that the several new hemoglobins I have described are only a few of the many, slightly different hemoglobins that, like those of the salmon, contribute to the complex, fine-tuned adaptation of diverse organisms to their environments. And these are but a tiny fraction of the "irreducibly complex" molecular adaptations to be found among vertebrates, insects, plants, and other forms of life. Behe, then, must be forced to see the designer's handiwork everywhere. Life must present him with countless instances of supernatural intervention—of miracles.

When scientists invoke miracles, they cease to practice science. Were a geologist to cite plate tectonics, a chemist hydrogen bonds, or a physicist gravity as an instance of the miraculous, he or she would be laughed out of the profession. Moreover, they would not be doing their job, which is to seek answers by posing and testing explanatory hypotheses. Faced with the unknown, as all scientists are, the scientist who invokes a miracle in effect says "this is unknowable" and admits defeat. It is only through confidence that the unknown is knowable that physical scientists have achieved explanation, and that biologists have advanced understanding of heredity, development, and evolution to heights scarcely hoped for just a few decades ago. Yet Behe, claiming a miracle in every molecule, would urge us to admit the defeat of reason, to despair of understanding, to rest content in ignorance. Even as biology daily grows in knowledge and insight, Behe counsels us to just give up.

Selection excerpted from:

Futuyma, Douglas J. 1997. Miracles and Molecules. *Boston Review* 22 (1): 29–30. Available from http://www.bostonreview.net/br22.1/futuyma.html.

MICRO/MACRO

The classic microevolutionary processes are natural selection, mutation, migration, and genetic drift, though some scientists would also include isolating mechanisms and other factors involved in speciation. These are genetically based mechanisms that affect gene pools of species, and that may result in change (adaptation) or stasis. Microevolutionary processes operate at the level of the species or population.

When speciation occurs, genetic material is no longer exchanged between groups of individuals, and each group begins an independent evolutionary history; this branching of species over and over through time results in the familiar "tree of life." Evolutionary biologists use the term "macroevolution" to refer to the topics relevant to understanding the distribution of patterns that emerge as species and lineages branche through time. Some of these are the rate of evolutionary change (rapid or slow), the pace of evolutionary change (gradual or jerky), adaptive radiation, morphological trends in lineages (e.g., whether body size gets smaller or larger), extinction or branching of a lineage, concepts (not covered in this book) such as species sorting, and the emergence of major new morphological features (such as segmentation, or shells, or the fusion or loss of bones). Scientists sometimes colloquially refer to macroevolution as "evolution above the species level," but this term does not do justice to the complexity of topics included within the concept.

Micro- and macroevolution are thus different levels of analysis of the same phenomenon: evolution. Macroevolution cannot solely be reduced to microevolution because it encompasses so many other phenomena: adaptive radiation, for example, cannot be reduced only to natural selection, though natural selection helps bring it about. Similarly, macroeconomics cannot be reduced to microeconomics: the food distribution system of the United States, involving import and export of goods from around the world, varied transportation systems and their regulation, multinational corporate decisions about supply and demand, investment and market control—and of course political decisions in Washington and in foreign nations—cannot be modeled by expanding the local organic farmers' market!

But the farmers' market is based on money exchange for food products consumed by people, which is also an important part of the national food distribution system; there are commonalities between micro- and macroeconomics. Similarly, the "currency" of microevolution is genes, and because evolution is a genealogical relationship of species, genes are highly relevant to understanding macroevolution—though the consideration of genes will not explain all macroevolutionary topics any more than the amount of money involved can explain macroeconomic systems.

Creationists' view of microevolution is similar to that of evolutionary biologists, but the two groups understand macroevolution very differently. Creationists accept microevolutionary processes affecting genetic variation of populations, and most also accept speciation, or the branching of a lineage into reproductively isolated groups. But creationists take literally the evolutionary biologists' definition of macroevolution as "evolution above the species level," and infer that major groups of living things such as phyla and classes—the upper taxonomic levels characterized by body plan differences—have a qualitatively different history than lower levels such as populations and species. They view the distinguishing features of phyla and classes as appearing suddenly, denying that such structures as segments, appendages, exoskeletons and the like could evolve through microevolutionary processes. Their definition of macroevolution thus overlaps only slightly with that of evolutionary biologists because they concentrate only on the emergence of new body plans or major features which distinguish "major kinds" of living things. Effectively, macroevolution to creationists equates to the inference of common ancestry, which they reject. Their view is that because God created living things as separate "kinds," major groups and the features distinguishing them could not have come about through natural processes, microevolutionary or otherwise. Their position is "micro yes, macro no."

There is a robust argument among evolutionary biologists over how new body plans or major new morphological features arose. No one disputes the importance of natural selection: it affects the genetic variation in populations, which may be the basis for a new species (in conjunction with isolating mechanisms). All parties likewise recognize the possibility or even likelihood of other biological mechanisms affecting morphological features that distinguish major groups of organisms. The issue in evolutionary biology is how and how much natural selection and other microevolutionary processes are supplemented by other mechanisms (such as regulatory genes operating early in embryological development).

Creationist Views of Micro/Macro

This first selection from the antievolutionist literature is from John Morris, current president of the Institute for Creation Research and the son of Henry M. Morris, the founder of modern Creation Science. He expresses quite well the essence of the Young Earth creationist view of the "micro/macro" issue.

What Is the Difference Between Macroevolution and Microevolution?

. . . There is much misinformation about these two words, and yet, understanding them is perhaps the crucial prerequisite for understanding the creation/evolution issue.

Macroevolution refers to major evolutionary changes over time, the origin of new types of organisms from previously existing, but different, ancestral types. Examples of this would be fish descending from an invertebrate animal, or whales descending from a land mammal. The evolutionary concept demands these bizarre changes.

Microevolution refers to varieties within a given type. Change happens within a group, but the descendant is clearly of the same type as the ancestor. This might better be called variation, or adaptation, but the changes are "horizontal" in effect, not "vertical." Such changes might be accomplished by "natural selection," in which a trait within the present variety is selected as the best for a given set of conditions, or accomplished by "artificial selection," such as when dog breeders produce a new breed of dog.

The small or microevolutionary changes occur by recombining existing genetic material within the group. As Gregor Mendel observed with his breeding studies on peas in the mid 1800's, there are natural limits to genetic change. A population of organisms can vary only so much. What causes macroevolutionary change?

Genetic mutations produce new genetic material, but do these lead to macroevolution? No truly useful mutations have ever been observed. The one most cited is the disease sickle-cell anemia, which provides an enhanced resistance to malaria. How could the occasionally deadly disease of SSA ever produce big-scale change?

Evolutionists assume that the small, horizontal microevolutionary changes (which are observed) lead to large, vertical macroevolutionary changes (which are never observed). This philosophical leap of faith lies at the core of evolution thinking.

A review of any biology textbook will include a discussion of microevolutionary changes. This list will include the variety of beak shape among the finches of the Galapagos Islands, Darwin's favorite example. Always mentioned is the peppered moth in England, a population of moths whose dominant color shifted during the Industrial Revolution, when soot covered the trees. Insect populations become resistant to DDT, and germs become resistant to antibiotics. While in each case, observed change was limited to microevolution, the inference is that these minor changes can be extrapolated over many generations to macroevolution.

In 1980 about 150 of the world's leading evolutionary theorists gathered at the University of Chicago for a conference entitled "Macroevolution." Their task: "to consider the mechanisms that underlie the origin of species" (Lewin, *Science*, vol. 210, pp. 883–887). "The central question of the Chicago conference was whether the mechanisms underlying microevolution can be extrapolated to explain the phenomena of macroevolution . . . the answer can be given as a clear, No."

Thus the scientific observations support the creation tenet that each basic type is separate and distinct from all others, and that while variation is inevitable, macroevolution does not and did not happen.

Selection excerpted from:

Morris, John D. 1996. What Is the Difference Between Macroevolution and Microevolution? *Back to Genesis* 94b, October.

Phillip Johnson is a major shaper of the Intelligent Design movement. In his book written for students, Defeating Darwinism, *he summarizes his view of the "micro/macro" issue, which is quite similar to that of John Morris.*

A Real Education in Evolution

. . . 2. *Learn to use terms precisely and consistently*. *Evolution* is a term of many meanings, and the meanings have a way of changing without notice. Dog breeding and finch-beak variations are frequently cited as typical examples of evolution. So is the fact that all the differing races of humans descend from a single parent, or even that Americans today are larger on average than they were a century ago (due to better nutrition). If relatively minor variations like that were all evolution were about, there would be no controversy, and even the strictest biblical fundamentalists would be evolutionists.

Of course evolution is about a lot more than in-species variation. The important issue is whether the dog breeding and finch-beak examples fairly illustrate the process that created animals in the first place. Using the single term *evolution* to cover both the controversial and the uncontroversial aspects of evolution is a recipe for misunderstanding.

At a minimum students must learn to distinguish between microevolution (cyclical variation within the type, as in the finch-beak example) and macroevolution (the vaguely described process that supposedly creates innovations such as new complex organs or new body parts). Don't be impressed by claims that in a few borderline cases microevolution may have produced, or almost produced, new "species." The definition of "species" is flexible and sometimes means no more than "isolated breeding group." By such a definition a fruit fly that breeds in August rather than June may be considered a new species, although it remains a fruit fly. The question is how we get insects and other basic groups in the first place. Darwinists typically (but not always) claim that macroevolution is just microevolution continued over a very long time. The claim is very controversial, and students should learn why.

3. *Keep your eye on the mechanism of evolution; it's the all-important thing*. . . . Darwin's mechanism was natural selection. Today, despite many efforts to find an alternative, there still isn't really a competitor to the two-part Darwinian mechanism of random variation (mutation) and natural selection. Darwinists argue with each other about the relative importance of chance and selection, but some combination of these two elements is just about the only game in town.

Remember that the mechanism has to be able to design and build very complex structures like wings and eyes and brains. Remember also that it has to have done this reliably again and again. Despite offhand references in the literature to possible alternatives, Darwinian natural selection remains the only serious candidate for a mechanism that might be able to do the job.

That, by the way explains why many Darwinists are reluctant to make a clear distinction between microevolution and macroevolution. They have evidence for a mechanism for minor variations, as illustrated by the finch-beak example, but have no distinct mechanism for the really creative kind of evolution, the kind that builds new body plans and new complex organs. Either macroevolution is just microevolution continued over a longer time, or it's a mysterious process with no known mechanism. A process like that isn't all that different from a miraculous or God-guided process, and it certainly wouldn't support those expansive philosophical statements about evolution being purposeless and undirected.

Selection excerpted from:

Johnson, Phillip E. 1997. *Defeating Darwinism by Opening Minds*. Downers Grove, IL: InterVarsity Press.

Evolutionary Biologists' View of Micro/Macro

Creationists have been quick to seize upon controversies within evolutionary biology about the relative importance of natural selection and other mechanisms of evolution, as evidence that evolutionary theory is weak. One controversy is over the pace of evolution: whether evolution tends to proceed slowly and gradually, or whether speciation takes place rapidly, with longer periods of little or no change between speciation events. The following selection discusses the controversy over "punctuated equilibria" and creationist application of it to the "micro/macro" question.

Debates About Macroevolution

[E]volutionary biologists would . . . argue that there is no basis for a distinction between the processes of microevolution and those of macroevolution. . . . [C]ontemporary neo-Darwinian orthodoxy is *evolutionary gradualism*. Speciation is regarded as a slow process in which many small genetic changes accumulate to result in the reproductive isolation of two populations. Population genetics studies these small genetic changes over short periods of time. Gradualists contend that one cannot admit the findings of population genetics without conceding the likelihood of major evolutionary changes by similar mechanisms. For large-scale evolution differs only in degree from the small modifications that are observed and studied. It is simply a matter of more changes, extended over a longer time.

Gradualist orthodoxy has been challenged, and Creationists have seized upon the words of the challengers. There are two independent challenges, that are frequently conflated by Creationists (and sometimes by biologists). The first challenge concerns the *tempo* of evolution. In a seminal paper, Niles Eldredge and Stephen Gould offered an alternative to the gradualist account of the pace of evolutionary change. Their proposal, the *punctuated-equilibrium model*, begins by building upon Mayr's theory of geographic speciation. Instead of thinking of species as evolving slowly and continually, Eldredge and Gould suggested that, in small isolated populations, change may be very rapid (in geological time). Stasis, the absence of change, is the norm for a species. Its central population is expected to persist relatively unmodified for millions of years. Detached from the central population, small peripheral groups can undergo morphological change very quickly, so that a period of isolation of a few thousand years may produce a population that is reproductively isolated and morphologically distinct from the parental stock. Evolution is not a slow, continuous process but a jerky affair. There are long pauses when nothing happens, interspersed with frenzied bursts of activity (p. 144).

. . . So far, the attack on gradualism has little bearing on the issue of whether Creationists can co-opt parts of evolutionary biology—specifically standard population genetics—without committing themselves to the possibility of the sorts of changes (evolution across "kinds") that they take to be impossible. The clash of ideas about the tempo of evolution does not . . . support the conclusion that the mechanisms of large-scale change are different from those that have been studied by population geneticists. Nevertheless, the debate already provides ammunition for Creationists to use.

Nobody has used the ammunition with more gusto than Duane Gish. The third edition of *Evolution? The Fossils Say No!* contains an extended discussion of current debates within evolutionary theory, designed to play defenders of the punctuated-equilibrium model against the gradualists. Here, for example, is Gish's discussion of a popular article by Gould . . . : "somewhat later in the same article Gould says: 'All paleontologists know that the fossil record contains precious little in the way of intermediate forms; transitions between major groups are characteristically abrupt'" (Gish 1979, 172). This citation, and others that are similar, should not mislead us into thinking that Creationist claims about the fossil record are upheld by reputable paleontologists.

Punctuated-equilibrium theorists do not deny that, *in one sense*, there are transitional forms between classes of organisms. They are happy to acknowledge that *Archeopteryx* is a transitional form between reptiles and birds, since it is an animal with some characteristics of both classes. . . . They are equally clear that the fossil record shows a sequence of therapsids becoming ever more mammal-like. What is denied is that there are *smooth* and *gradual* transitions among *species*. Focusing on the relatively small differences between related species, punctuated-equilibrium theorists point out that we do not find these differences bridged by a continuous sequence of intermediates. However, to deny the existence of intermediates on a fine scale is not to suggest that we never encounter intermediates between large groups of organisms, creatures that exhibit fully developed features of the later group while retaining some ancestral characteristics (pp. 145–148).

. . . Evolutionary theorists often agree that two particular species evolved from a common ancestor, even though they disagree about how the evolution occurred. One theorist may suggest that the process involved a succession of genetic changes with small phenotypic effect. The other may insist that the transition required mutations affecting developmental patterns, so that small genetic alterations were amplified in the phenotype. It is illegitimate to deploy the arguments offered by both sides, arguments that criticize the rival accounts of the *process* of evolution, to cast doubt on the *existence* of evolution. *For we can know that a species is related to an ancestral population by evolutionary descent, even though the details of the transition are controversial.* [To do so] . . . is simply to ignore all the reasons that scientists may have for recognizing an evolutionary relationship, such as intricate similarities in morphology.

. . . The suggestion that macroevolution should be divorced from microevolution provides Creationists only with a debating point. It allows Creationists to say that there are some evolutionary theorists who distinguish the mechanisms studied in classical population genetics from those they take to be involved in large-scale evolutionary change (or, more exactly, in some cases of large-scale evolutionary change). But this is not to suppose that the distinction drawn by heterodox evolutionists is that favored by the Creationists (p. 150).

[References omitted.]

Selection excerpted from:

Kitcher, Philip. 1982. *Abusing Science: The Case Against Creationism*. Cambridge, MA: MIT Press.

The previous selection explained that the punctuated equilibrium controversy is concerned with the tempo or pace of evolutionary change, and that there are also disputes taking place in evolutionary biology about the mode of change: whether microevolutionary processes alone can explain major changes in body plan. The next author, the evolutionary biologist Douglas Futuyma, contends that microevolutionary processes, specifically natural selection, can explain many differences that distinguish major groups—what creationists think of as "macroevolution." But natural selection is not the only microevolutionary process operating on populations, as will be seen in the last selection in this series.

Creationist Arguments

. . . 4. Creationists, however, deny that mutation, recombination, and natural selection can form new, complex features.

. . . It is not true, however, that mutations are almost universally harmful. Whether mutations that alter the metabolic abilities of bacteria, confer insecticide resistance on a fly, or change the height and growth form of a plant are harmful or beneficial depends on the environment. Evolutionary theory does not postulate that "mutations must be primarily beneficial"—only that some are. Put a culture of bacteria, fungi, or flies into a novel environment and within a few generations it will have evolved improved adaptation, even if, as is easily done with these organisms, you begin with a population of genetically identical individuals, and even if the majority of the mutations in that population are negative (p. 185).

. . . The crux of the creationist objection, though, lies in the emphasis on "real novelties." Creationists have responded to the fact that biologists have observed genetic change in organisms by inventing the idea that each "kind" was created with a great variety of genes. However, "Modern molecular biology, with its penetrating insight into the remarkable genetic code, has further confirmed that normal variations operate only within the range specified by the DNA for the particular type of organism, so that no truly novel characteristics, producing higher degrees of order or complexity, can appear" (Morris, 1974: 51).

Modern molecular biology has confirmed no such thing. It has confirmed that mutations can affect a small or a large part of a gene or of a chromosome; that new genetic information can come into existence by the duplication of preexisting genes and by exchanges of nucleotides to form entirely new gene sequences; that mutations can alter the organism's biochemistry a great deal or not at all. Together with the study of development, molecular genetics has shown that even slight genetic changes can provide enzymes with new biochemical functions; can alter the size, shape, and growth rate of every feature of an organism's body; and can produce changes much like those that distinguish different, related species. The "range specified by the DNA for the particular type of organism" is a creationist fiction for which none of molecular biology offers support.

The "higher degrees of order or complexity" that creationists believe cannot evolve are actually impossible to define. Begin with a reptile, for example, and imagine one of the lower jaw bones becoming larger and the other smaller, so that they finally are disconnected. Is this an increase in complexity? It is one of the chief defining features of the class Mammalia. The single-cusp tooth of the reptiles develops small accessory cusps. Is this a higher degree of complexity? The different variations on the multicusp theme are the basis of much of the adaptive radiation of the mammals into different ways of life, and genetic variations in tooth form are common within many species of mammals. Imagine slight variations in the position of the eyes, from the side of the head toward the front. Such variations in shape and orientation are characteristic of almost every feature of organisms, although this particular one is a major adaptive feature of the primates. Is it really more complex than similar ones in "lesser" species? The "higher degrees of order and complexity" that so impress the creationists are, in a sense, illusory. The "complexity" of a horse or a dandelion is actually just a collection of individual features. Each of these can (and usually did) evolve independently, and each is not very drastic remodeling of the features of the ancestor. And the material for remodeling is evident in the variation within species.

The creationists continue to argue that variation cannot transcend the limits of the "kind"—a Biblical term that has no meaning in modern taxonomy. Yet they have no idea how to define or recognize a "kind." Are lizards and snakes different "kinds" because iguanas are so different from cobras, or the same "kind" because there are so many intermediate snakelike lizards and lizardlike snakes? This vagueness is convenient for the creationist argument, of course, because whenever a biologist or paleontologist finds an intermediate between two "kinds," the creationist can claim that they are the same "kind" after all. The argument that genetic changes cannot bring about new, more complex "kinds" of organisms rests on the belief that organisms fall into discrete, higher and lower "kinds." But they do not (pp. 186–187).

REFERENCE CITED

Morris, Henry M., ed. 1974. *Scientific Creationism.* San Diego: Creation-Life Publishers.

Selection excerpted from:

Futuyma, Douglas J. 1995. *Science on Trial: The Case for Evolution.* Sunderland, MA: Sinauer Associates.

The following article proposes that some other "macroevolutionary" (body plan) differences may be produced by genetic factors operating during embryological growth and development. The new field of "evo-devo" (see chapters 2 and 6) explores these and other subjects. The review article by Hughes and Kaufman introduces some of these ideas as applied to the arthropod body plan. The full article includes many references to original research, and a comprehensive summary of research to date (2002).

Hox Genes and the Evolution of the Arthropod Body Plan

Introduction

The evolution of different animal body plans is one of the great mysteries of biology. The phylum Arthropoda, for instance, includes millions of extinct and extant species with diverse morphology, including ticks and trilobites, crabs and centipedes, spiders, shrimps, and spittlebugs. From a basic organization consisting of a series of segments encased in an exoskeleton, the various arthropod groups have developed a multitude of specialized forms. Such morphological diversity begs the question of how it arose. To answer this question by invoking natural selection is correct—but insufficient. The fangs of a centipede, the sucking proboscis of a bug, and the claws of a lobster indisputably accord these organisms a fitness advantage. However, the crux of the mystery is this: From what developmental genetic changes did these novelties arise in the first place? To begin to address this question, researchers have turned to the Hox genes.

The Hox genes are a set of related genes encoding homeodomain transcription factors. These genes are important developmental regulators, acting together to determine the identity of segments along the anterior-posterior axis of the embryo. Each Hox gene is thought to control the expression of a variety of target genes, which may number into the hundreds. Thus the activity of a single Hox gene can be sufficient to induce an entire "developmental module" of target genes, which act in concert to confer a particular identity upon a developing segment. Express a Hox gene in the wrong place, and a completely different kind of appendage will develop. For example, gain-of-function mutations in *antennapedia* cause legs instead of antennae to grow out of the head of the fruit fly. Because of their important role in determining segment identity, the Hox genes are closely associated with development of the regionalization of the body plan.

In fact, it seems to be the expression profile of the Hox genes and their activity that are largely responsible for determining the body patterning of a species. Therefore, evolutionary changes in the expression of Hox genes or their activity might have caused evolutionary changes in body patterning. Why does the lobster have two pairs of specialized front legs (maxillipeds), whereas brine shrimp have none? It seems to be correlated with a shift in Hox gene expression. Why do insects have a thorax and abdomen, whereas centipedes have just one long homonomous trunk? Again, the difference may be due to the different expression of Hox genes.

The Hox genes are typically found together in a single complex on the chromosome. The genes promote the identity of segments along the anterior-posterior axis of the embryo in the same order in which they lie on the chromosome. . . . In addition to conferring identity to their own segment(s) via target genes, most Hox genes interact with each other to maintain proper expression domain boundaries by both positive and negative regulation. For instance, in general a more posterior gene will suppress the expression or the function of the more anterior Hox gene, phenomena known as posterior prevalence, posterior dominance, or phenotypic suppression. Thus, Hox genes are not merely located together on the chromosome, but they also interact together in complex ways to collectively define segment identity along the anterior-posterior axis of the embryo. . . .

Our goal in this review is to introduce the reader to the role of the Hox genes in the evolution of the arthropods. The cast of characters includes some genes that have changed their developmental roles wildly, some genes that have merely tweaked their expression patterns in different species, and some that have stubbornly remained expressed in a nearly invariant pattern.

. . . Because of the high conservation of the Hox genes and their important role in specifying segment identity, they have been studied in many arthropod species. Unfortunately, this wealth of information is scattered among dozens of original data articles and thus is not easily compared.

. . . After all, in general researchers are not obsessed with studying, say, a particular species of centipede purely for its own sake. Rather, each seemingly arcane study is meant to provide some enlightenment into general principles of evolution that may apply to all animal life. But have we actually achieved this quixotic goal? To some degree the answer is yes. In fact, the study of changes in arthropod Hox genes might well be called one of the first success stories of the nascent field of evo-devo. In addition to a greater understanding of the development of particular arthropod species, emerging themes have led to some provocative general models for how Hox genes may have been involved in evolution. . . . The comparative work in arthropod Hox genes has been truly revolutionary because it has succeeded in providing some of the first concrete models of the mechanistic basis of morphological evolution (pp. 459–462).

. . . Although our understanding of genetic events that occurred millions of years ago can never be totally conclusive, the ability to simulate evolutionary events in the laboratory environment by manipulation of a critical gene would be convincing evidence to support a theoretical model. With the increased power of expanding functional techniques, the field of arthropod evo-devo is coming to the stage of its development in which some of the beautiful theories described here are bound to be shattered by some ugly facts. But as developmental biologists know, although the coming-of-age process may be awkward at times, it is a necessary step to full maturity. Finally, in further studies we must avoid Hox snobbery. Although Hox genes are important developmental regulatory genes and have been extremely practical to begin studying the evolution of the arthropod, analysis of the *Drosophila* genome suggests that there are approximately 13,590 other genes in the genome, which we have barely begun to explore in other species (p. 494).

[References omitted.]

Selection excerpted from:

Hughes, Cynthia L., and Thomas G. Kaufman. 2002. Hox Genes and the Evolution of the Arthropod Body Plan. *Evolution and Development* 4 (6): 459–499.

CHAPTER 9

Legal Issues

INTRODUCTION

To understand legal issues involving the creation and evolution controversy, one must first understand the three clauses of the First Amendment of the United States Constitution. The Religion Clause states, "Congress shall make no law respecting an establishment of religion, or prohibiting the free exercise thereof." The Free Speech Clause states, "or abridging the freedom of speech, or of the press." The third clause proclaims "the right of the people peaceably to assemble, and to petition the government for a redress of grievances." All of the legal decisions generated by the creation and evolution controversy have been decided based on interpretations of the Religion and Free Speech clauses of the First Amendment.

The *Establishment* and *Free Exercise* clauses, taken together, mean that public institutions have to be religiously neutral. Public schools, for example, can neither advance nor inhibit religion. Neutrality has meant different things to different people: to William Jennings Bryan and Creation Science proponents during the 1970s and 1980s (see chapter 5), neutrality meant not teaching evolution, which they viewed as a religious perspective, or, if evolution were taught, it should be "balanced" with the teaching of creationism. To opponents of creationism, neutrality means teaching only scientific views—keeping sectarian religion out of the classroom.

Laws have been judged for Establishment Clause constitutionality by applying a three-part test devised in a 1971 Supreme Court case, *Lemon v. Kurtzman*. The *Lemon* test requires that a bill or practice must have a secular rather than a religious *purpose*; it must not have an *effect* that either promotes or inhibits religion, and it must not create undue *entanglement* between government and religion. Failure on any of the three "prongs" of *Lemon* means the bill is unconstitutional. All of the creationism cases decided after 1971 have referred to the *Lemon* test, and all of them have been struck down on at least the first—purpose—prong.

Tension exists between the Establishment and the Free Exercise clauses. An early (if unsuccessful) legal strategy of antievolutionists was to claim that the teaching of evolution violated a child's free exercise of religion because teaching evolution supposedly was an attack on religious belief (Larson 2003: 131). On the other hand, presenting creationism in science class—encouraged by antievolutionists as a means to counter the allegedly antireligious effect of teaching evolution—violates the Establishment Clause. The Free Speech Clause has also been invoked in support of a teacher's right to teach evolution by both John Scopes (see chapter 5) and by Susan Epperson (see below). More recently, creationist teachers have argued a free speech/academic freedom "right" to add creationism to the curriculum (see *Webster*, below), but courts have held that teachers at the kindergarten–grade 12 level (K–12) must follow the curriculum set by the district. Precollege teachers have far less academic freedom than do university professors; hence the free speech argument has not fared well in court.

When it comes to religion in schools, courts since the 1980s have come down more strongly on the Establishment Clause side of the First Amendment than on other clauses. Because attendance in public schools is mandatory, and because parents have the right to guide their children's religious views, courts have firmly restricted teachers from proselytizing students. For example, teachers cannot lead prayers or religious after-school clubs. Because creationism is an inherently religious concept, advocating it—whether in biblical or Creation Science form—is considered proselytization, and is therefore unconstitutional.

The legal history of the creation and evolution controversy can be divided into three parts, and this is reflected in the organization of readings in this chapter. The first period is that of attempts to ban evolution; second is the period of "equal time" for evolution and Creation Science, and finally, the Neocreationism period extends to the present day. Chapters 5 and 6 present context for these selections.

• •

BANNING EVOLUTION

The unidimensional public view of William Jennings Bryan as a religious buffoon, perhaps stimulated by association with the unattractive prosecutor in the Scopes Trial–like play *Inherit the Wind*, is unfortunate and unfair. Bryan was a much more complex figure than he is usually given credit for being, and his campaigns for social justice played an important role in producing a more humane workplace, among other issues. His views on evolution, reflecting a common view that evolution was not only antireligious but also a path to social evil (see chapter 4) helped bring about laws banning the teaching of evolution.

William Jennings Bryan on Evolution

. . . I object to the theory for several reasons. First, it is a dangerous theory. If a man link himself in generations with the monkey, it then becomes an important question whether he is going towards him or coming from him—and I have seen them going in both directions. I don't know of any argument that can be used to prove that man is an improved monkey that may not be used just as well to prove that the monkey is a degenerate man, and the latter theory is more plausible than the former . . . ("The Prince of Peace," p. 41).

Go back as far as we may, we cannot escape from the creative act, and it is just as easy for me to believe that God created man as he is as to believe that, millions of years ago, He created a germ of life and endowed it with power to develop into all that we see to-day. I object to the Darwinian theory, until more conclusive proof is produced, because I fear we shall lose the consciousness of God's presence in our daily life, if we must accept the theory that through all the ages no spiritual force has touched the life of man or shaped the destiny of nations. But there is another objection. The Darwinian theory represents man as reaching his present perfection by the operation of the law of hate—the merciless law by which the strong crowd out and kill off the weak. . . . I prefer to believe that love rather than hatred is the law of development . . . ("The Prince of Peace," p. 42).

Our first objection to Darwinism is that it is not true. I may add here, so I will not have to refer to it again, that I am answering theistic evolution as well as atheistic evolution; I do not make any difference between them . . . for the theistic evolutionist and the atheistic evolutionist walk along hand in hand until they reach the beginning of life. They are nearer together than either of them is to the Christian . . . ("Is the Bible True?" p. 82).

Let us look at some of their guesses. . . . Do you know how the eye came? . . . They guess that an animal that did not have any eyes, away back yonder had a piece of pigment or freckle on the skin—it just happened—and when the sun's rays were traveling over the animal's body and came to that piece of pigment or freckle, they converged there more than elsewhere and that made it warmer there than elsewhere, and that irritated the skin there instead of elsewhere, and that brought a nerve there instead of somewhere else, and the nerve developed into an eye. And then another freckle, and another eye; in the right place and at the right time. Can you beat it? ("Is the Bible True?" pp. 82–83).

Selection excerpted from:

Cornelius, R. M., ed. 1996. *Selected Orations of William Jennings Bryan: The Cross of Gold Centennial Edition*. Dayton, TN: William Jennings Bryan College. Used by permission.

Epperson v. Arkansas (1968)

. . . The Arkansas law makes it unlawful for a teacher in any state-supported school or university "to teach the theory or doctrine that mankind ascended or descended from a lower order of animals," or "to adopt or use in any such institution a textbook that teaches" this theory. Violation is a misdemeanor and subjects the violator to dismissal from his position. . . .

[T]he law must be stricken because of its conflict with the constitutional prohibition of state laws respecting an establishment of religion or prohibiting the free exercise thereof. The overriding fact is that Arkansas' law selects from the body of knowledge a particular segment which it proscribes for the sole reason that it is deemed to conflict with a particular religious doctrine; that is, with a particular interpretation of the Book of Genesis by a particular religious group. . . .

Government in our democracy, state and national, must be neutral in matters of religious theory, doctrine, and practice. It may not be hostile to any religion or to the advocacy of no-religion; and it may not aid, foster, or promote one religion or religious theory against another or even against the militant opposite. The First Amendment mandates governmental neutrality between religion and religion, and between religion and nonreligion. . . .

While study of religions and of the Bible from a literary and historic viewpoint, presented objectively as part of a secular program of education, need not collide with the First Amendment's prohibition, the State may not adopt programs or practices in its public schools or colleges which "aid or oppose" any religion. This prohibition is absolute. It forbids alike the preference of a religious doctrine or the prohibition of theory which is deemed antagonistic to a particular dogma. As Mr. Justice Clark stated in *Joseph Burstyn, Inc. v Wilson*, "the state has no legitimate interest in protecting any or all religions from views distasteful to them. . . . "

Arkansas' law cannot be defended as an act of religious neutrality. Arkansas did not seek to excise from the curricula of its schools and universities all discussion of the origin of man. The law's effort was confined to an attempt to blot out a particular theory because of its supposed conflict with the Biblical account, literally read. Plainly, the law is contrary to the mandate of the First, and in violation of the Fourteenth, Amendment to the Constitution.

[Internal citations deleted.]

Selection excerpted from:

Epperson v. Arkansas 393 U.S. 97 (1968).

• •

EQUAL TIME FOR CREATION SCIENCE

In 1981, Arkansas was the first state to pass an "equal time for Creation Science and evolution" law. This case is the Federal District Court decision that struck it down. As discussed in chapter 5, the *McLean* decision was not appealed, so its conclusions were not extended beyond its district. However, its reasoning was highly influential in the subsequent Supreme Court case that struck down "equal time" laws nationwide.

McLean v. Arkansas Board of Education

On March 19, 1981, the Governor of Arkansas signed into law Act 590 of 1981, entitled "Balanced Treatment for Creation-Science and Evolution-Science Act." . . . Its essential mandate is stated in its first sentence: "Public schools within this State shall give balanced treatment to creation-science and to evolution-science." On May 27, 1981, this suit was filed challenging the constitutional validity of Act 590 on three distinct grounds.

First, it is contended that Act 590 constitutes an establishment of religion prohibited by the First Amendment to the Constitution, which is made applicable to the states by the Fourteenth Amendment. Second, the plaintiffs argue the Act violates a right to academic freedom which they say is guaranteed to students and teachers by the Free Speech Clause of the First Amendment. Third, plaintiffs allege the Act is impermissibly vague and thereby violates the Due Process Clause of the Fourteenth Amendment.

The individual plaintiffs include the resident Arkansas Bishops of the United Methodist, Episcopal, Roman Catholic and African Methodist Episcopal Churches, the principal official of the Presbyterian Churches in Arkansas, other United Methodist, Southern Baptist and Presbyterian clergy, as well as several persons who sue as parents and next friends of minor children attending Arkansas public schools. One plaintiff is a high school biology teacher. All are also Arkansas taxpayers. Among the organizational plaintiffs are the American Jewish Congress, the Union of American Hebrew Congregations, the American Jewish Committee, the Arkansas Education Association, the National Association of Biology Teachers and the national Coalition for Public Education and Religious Liberty, all of which sue on behalf of members living in Arkansas.

The defendants include the Arkansas Board of Education and its members, the Director of the Department of Education, and the State Textbooks and Instructional Materials Selecting Committee. . . .

The unusual circumstances surrounding the passage of Act 590, as well as the substantive law of the First Amendment, warrant an inquiry into the stated legislative purposes. The author of the Act has publicly proclaimed the sectarian purpose of the proposal. The Arkansas residents who sought legislative sponsorship of the bill did so for a purely sectarian purpose. These circumstances alone may not be particularly persuasive, but when considered with the publicly announced motives of the legislative sponsor made contemporaneously with the legislative process; the lack of any legislative investigation, debate or consultation with any educators or scientists; the unprecedented intrusion in school curriculum; and official history of the State of Arkansas on the subject, it is obvious that the statement of purpose has little, if any, support in fact. The State failed to produce any evidence which would warrant an inference or conclusion that at any point in the process anyone considered the legitimate educational value of the Act. It was simply and purely an effort to introduce the Biblical version of creation into the public school curricula. The only inference which can be drawn from these circumstances is that the Act was passed with the specific purpose by the General Assembly of advancing religion. The Act therefore fails the first prong of the three-pronged test, that of secular legislative purpose, as articulated in *Lemon v. Kurtzman*. . . .

The evidence establishes that the definition of "creation science" contained in 4(a) has as its unmentioned reference the first 11 chapters of the Book of Genesis. Among the many creation epics in human history, the account of sudden creation from nothing, or creation ex nihilo, and subsequent destruction of the world by flood is unique to Genesis. The concepts of 4(a) are the literal Fundamentalists' view of Genesis. Section 4(a) is unquestionably a statement of religion, with the exception of 4(a)(2), which is a negative thrust aimed at what the creationists understand to be the theory of evolution.

Both the concepts and wording of Section 4(a) convey an inescapable religiosity. Section 4(a)(1) describes "sudden creation of the universe, energy and life from nothing." Every theologian who testified, including defense witnesses, expressed the opinion that the statement referred to a supernatural creation which was performed by God. . . .

The facts that creation-science is inspired by the Book of Genesis and that Section 4(a) is consistent with a literal interpretation of Genesis leave no doubt that a major effect of the Act is the advancement of particular religious beliefs. The legal impact of this conclusion will be discussed further at the conclusion of the Court's evaluation of the scientific merit of creation-science.

The approach to teaching "creation-science" and "evolution-science" found in Act 590 is identical to the two-model approach espoused by the Institute for Creation Research and is taken almost verbatim from ICR writings. It is an extension of Fundamentalists' view that one must either accept the literal interpretation of Genesis or else believe in the godless system of evolution.

The two-model approach of the creationists is simply a contrived dualism which has no scientific factual basis or legitimate educational purpose. It assumes only two explanations for the origins of life and existence of man, plants and animals: it was either the work of a creator or it was not. Application of these two models, according to creationists, and the defendants, dictates that all scientific evidence which fails to support the theory of evolution is necessarily scientific evidence in support of creationism and is, therefore, creation-science "evidence" in support of Section 4(a). . . .

The methodology employed by creationists is another factor which is indicative that their work is not science. A scientific theory must be tentative and always subject to revision or abandonment in light of facts that are inconsistent with, or falsify, the theory. A theory that is by its own terms dogmatic, absolutist, and never subject to revision is not a scientific theory.

The creationists' methods do not take data, weigh it against the opposing scientific data, and thereafter reach the conclusions stated in Section 4(a). Instead, they take the literal wording of the Book of Genesis and attempt to find scientific support for it. . . .

Implementation of Act 590 will have serious and untoward consequences for students, particularly those planning to attend college. Evolution is the cornerstone of modern biology, and many courses in public schools contain subject matter relating to such varied topics as the age of the earth, geology and relationships among living things. Any student who is deprived of instruction as to the prevailing scientific thought on these topics will be denied a significant part of science education. Such a deprivation through the high school level would undoubtedly have an impact upon the quality of education in the State's colleges and universities, especially including the pre-professional and professional programs in the health sciences. . . .

The defendants presented Dr. Larry Parker, a specialist in devising curricula for public schools. He testified that the public school's curriculum should reflect the subjects the public wants in schools. The witness said that polls indicated a significant majority of the American public thought creation-science should be taught if evolution was taught. . . .

The application and content of First Amendment principles are not determined by public opinion polls or by a majority vote. Whether the proponents of Act 590 constitute the majority or the minority is quite irrelevant under a constitutional system of government. No group, no matter how large or small, may use the organs of government, of which the public schools are the most conspicuous and influential, to foist its religious beliefs on others.

[Internal citations deleted.]

Selection excerpted from:

McLean v. Arkansas Board of Education 529 F. Supp. 1255 (E.D. La. 1982).

Edwards *is the Supreme Court decision discussed in chapter 5 that struck down "equal time" laws nationally.*

Edwards v. Aguillard (1987)

Louisiana's "Creationism Act" forbids the teaching of the theory of evolution in public elementary and secondary schools unless accompanied by instruction in the theory of "creation science." The Act does not require the teaching of either theory unless the other is taught. It defines the theories as "the scientific evidences for [creation or evolution] and inferences from those scientific evidences. . . . "

Held:

1. The Act is facially invalid as violative of the Establishment Clause of the First Amendment, because it lacks a clear secular purpose.

(a) The Act does not further its stated secular purpose of "protecting academic freedom." It does not enhance the freedom of teachers to teach what they choose and fails to further the goal of "teaching all of the evidence." Forbidding the teaching of evolution when creation science is not also taught undermines the provision of a comprehensive scientific education. Moreover, requiring the teaching of creation science with evolution does not give schoolteachers a flexibility that they did not already possess to supplant the present science curriculum with the presentation of theories, besides evolution, about the origin of life. Furthermore, the contention that the Act furthers a "basic concept of fairness" by requiring the teaching of all of the evidence on the subject is without merit. Indeed, the Act evinces a discriminatory preference for the teaching of creation science and against the teaching of evolution by requiring that curriculum guides be developed and resource services supplied for teaching creationism but not for teaching evolution, by limiting membership on the resource services panel to "creation scientists," and by forbidding school boards to discriminate against anyone who "chooses to be a creation-scientist" or to teach creation science, while failing to protect those who choose to teach other theories or who refuse to teach creation science. A law intended to maximize the comprehensiveness and effectiveness of science instruction would encourage the teaching of all scientific theories about human origins. Instead, this Act has the distinctly different purpose of discrediting evolution by counter-balancing its teaching at every turn with the teaching of creationism.

(b) The Act impermissibly endorses religion by advancing the religious belief that a supernatural being created humankind. The legislative history demonstrates that the term "creation science," as contemplated by the state legislature, embraces this religious teaching. The Act's primary purpose was to change the public school science curriculum to provide persuasive advantage to a particular religious doctrine that rejects the factual basis of evolution in its entirety. Thus, the Act is designed either to promote the theory of creation science that embodies a particular religious tenet or to prohibit the teaching of a scientific theory disfavored by certain religious sects. In either case, the Act violates the First Amendment. . . .

The Court has been particularly vigilant in monitoring compliance with the Establishment Clause in elementary and secondary schools. Families entrust public schools with the education of their children, but condition their trust on the understanding that the classroom will not purposely be used to advance religious views that may conflict with the private beliefs of the student and his or her family. Students in such institutions are impressionable and their attendance is involuntary. The State exerts great authority and coercive power through mandatory attendance requirements, and because of the students' emulation of teachers as role models and the children's susceptibility to peer pressure. Furthermore, "[t]he public school is at once the symbol of our democracy and the most pervasive means for promoting our common destiny. In no activity of the State is it more vital to keep out divisive forces than in its schools. . . . "

Consequently, the Court has been required often to invalidate statutes which advance religion in public elementary and secondary schools.

Therefore, in employing the three-pronged *Lemon* test, we must do so mindful of the particular concerns that arise in the context of public elementary and secondary schools. . . .

We do not imply that a legislature could never require that scientific critiques of prevailing scientific theories be taught. Indeed, the Court acknowledged in *Stone* that its decision forbidding the posting of the Ten Commandments did not mean that no use could ever be made of the Ten Commandments, or that the Ten Commandments played an exclusively religious role in the history of Western Civilization. In a similar way, teaching a variety of scientific theories about the origins of humankind to schoolchildren might be validly done with the clear secular intent of enhancing the effectiveness of science instruction. But because the primary purpose of the Creationism Act is to endorse a particular religious doctrine, the Act furthers religion in violation of the Establishment Clause. . . .

The Louisiana Creationism Act advances a religious doctrine by requiring either the banishment of the theory of evolution from public school classrooms or the presentation of a religious viewpoint that rejects evolution in its entirety. The Act violates the Establishment Clause of the First Amendment because it seeks to employ the symbolic and financial support of government to achieve a religious purpose. The judgment of the Court of Appeals therefore is Affirmed.

[Internal citations deleted.]

Selection excerpted from:

Edwards v. Aguillard 482 U.S. 578 (1987).

● ●

NEOCREATIONISM

Recall from chapter 6 that present-day antievolutionists—aware of the above court decisions, and especially the *purpose* prong of *Lemon*—try to avoid reference to creationism or religion. Instead, evolution is declared a *scientifically* controversial topic and legislation is proposed that would require teachers to "give students all the evidence," which means to include alleged weaknesses as well as strengths of evolution. By appealing to ingrained American cultural standards of fairness, such arguments find a receptive audience.

An example of this more subtle approach is an amendment proposed by Senator Rick Santorum to the 2002 "No Child Left Behind" education bill. Although couched as a "critical thinking" statement, only evolution is singled out from all potentially controversial scientific theories. Even though the amendment failed and appears only in altered form in the conference committee report, language from the amendment is showing up in antievolution legislation around the country.

The Santorum Amendment and Its Repercussions

The Original Amendment

It is the sense of the Senate that

(1) good science education should prepare students to distinguish the data or testable theories of science from philosophical or religious claims that are made in the name of science; and
(2) where biological evolution is taught, the curriculum should help students to understand why the subject generates so much continuing controversy, and should prepare the students to be informed participants in public discussions regarding the subject.

Item 78, Conference Committee Report

The conferees recognize that a quality science education should prepare students to distinguish the data and testable theories of science from religious or philosophical claims that are made in the name of science. Where topics are taught that may generate controversy (such as biological evolution), the curriculum should help students to understand the full range of scientific views that exist, why such topics may generate controversy, and how scientific discoveries can profoundly affect society.

Legislation Inspired by the Santorum Language

Georgia HB 1563 (2002)

. . . In recognition of the fact that a quality science education should prepare students to distinguish the data and testable theories of science from philosophical claims that are made in the name of science, the State Board of Education is authorized to promulgate rules and regulations and develop a curriculum for topics that may generate controversy, such as biological evolution, to help students understand the full range of scientific views that exist, why such topics may generate controversy, and how scientific discoveries can profoundly affect society. (Died in committee.)

Ohio HB 481 (2002)

A) Encourage the presentation of scientific evidence regarding the origins of life and its diversity objectively and without religious, naturalistic, or philosophic bias or assumption;

B) Require that whenever explanations regarding the origins of life are presented, appropriate explanation and disclosure shall be provided regarding the historical nature of origins science and the use of any material assumption which may have provided a basis for the explanation being presented;

C) Encourage the development of curriculum that will help students think critically, understand the full range of scientific views that exist regarding the origins of life, and understand why origins science may generate controversy. (Died in committee.)

"Scientific Alternatives to Evolution": Intelligent Design

One alleged scientific alternative to evolution is Intelligent Design theory, and, reminiscent of the "equal time for creation science" efforts of the late 1970s and 1980s, attempts to mandate the teaching of ID are increasing. In at least one case, ID was linked with the Santorum conference committee language when the Phoenixville, Pennsylvania, school district vice-president argued that the Santorum language encouraged students "to have classroom discussions of alternatives to the theory of evolution, including one called 'intelligent design,' which says that life is so complex that it must have been initiated by some higher, intelligent power" (Hardy 2002). The issue has also appeared at the state legislative level.

Michigan HB 4382 (2001), HB 4946 (2003)

In the science standards for middle school and high school, all references to "evolution" and "natural selection" shall be modified to indicate that these are unproven theories by adding the phrase "describe how life may be the result of the purposeful, intelligent design of a Creator." (Died in committee in 2001; reintroduced in 2003.)

Darwin Faces a New Rival; A Roseville High School Parent Urges That "Intelligent Design" Also Be Taught in Biology

A parent's request that Roseville high schools teach ideas that rebut Darwin's theory of evolution could set the stage for debate over what critics call the newest version of creationism.

When Roseville Joint Union High School District trustees took the first step toward approving a new biology textbook earlier this month, parent Larry Caldwell asked that supplementary materials be taught in conjunction with the text, which, like most biology books, presents the theory of evolution to explain the origins of life.

"Evolution doesn't represent all the views in the scientific community and probably doesn't represent the best views," said Caldwell, who has a child at Granite Bay High School. "Rather than giving our students the most up-to-date view of science, we're giving an outdated view that reflects the opinions of only one group of scientists."

Caldwell said he would like to work with district officials in gathering educational materials that present a theory called "intelligent design." District officials said they will assemble and review the materials in the coming months, seeking input from science teachers. . . .

Selection excerpted from:

Rosen, Laurel. 2003. Darwin Faces a New Rival; A Roseville High School Parent Urges That "Intelligent Design" Also Be Taught in Biology. *Sacramento Bee*, June 22, pp. B1, B3.

Evidence Against Evolution

Both ID and Creation Science promoters encourage the teaching of "evidence for and against evolution." A series of bills with similar wording got their start with a 1996 Ohio bill that was stimulated by the grassroots efforts of a retired Wisconsin teacher. John Hansen founded Operation T.E.A.C.H.E.S. (Teach Evolution Accurately, Consistently, Honestly, Equitably, Scientifically) and drove around the country between 1995 and 2000, trying to talk state legislators in Wisconsin, Minnesota, Indiana, Iowa, Ohio, Kentucky, Alaska, Georgia, and New Mexico into sponsoring his model bill (Hansen 1997, 1999, 2000). Ohio State Representative Ron Hood formatted Hanson's idea as legislation (Trevas 1996) and introduced it (without success) in 1996 and 2000. A Georgia legislator also introduced the "Hood Bill," and when Hansen retired to Arizona, he persuaded an Arizona legislator to introduce the bill as well. None of these bills passed.

Hansen claimed in his newsletter that legislators in Alaska, New Mexico, and Kentucky also introduced his legislation, but NCSE has no record of this.

Ohio HB 62 (1996) and HB 679 (2000)

Whenever a theory of the origin of humans or other living things that might commonly be referred to as "evolution" is included in the instructional program provided by any school district or educational service center, both scientific evidence and related arguments supporting or consistent with the theory and scientific evidence and related arguments problematic for, inconsistent with, or not supporting the theory shall be included.

Georgia HB 1133 (1998)

Whenever a theory of the origin of humans or other living things that might commonly be referred to as "evolution" is included in a course of study offered by a local unit of administration, both scientific evidence supporting or consistent with the theory and scientific evidence problematic for, inconsistent with, or not supporting the theory shall be included.

Arizona HB 2585 (2000)

If instruction is provided on the theory of evolution concerning the origin of human beings and other living organisms, the teacher shall present scientific evidence that supports or is consistent with the theory of evolution and scientific evidence that does not support or is not consistent with the theory of evolution.

Ohio Science Standards (2002)

In 2002, after failing to get the Ohio State Board of Education to include Intelligent Design in the state science standards, antievolutionists shifted tactics to persuade friendly school board members to include a measure that they could argue was a requirement for teaching "evidence against evolution." The Ohio Science Education Standards for Life Sciences at the tenth grade level included Standard 23, which read: "Describe how scientists today continue to investigate and critically analyze aspects of evolutionary theory." Although scientists routinely critically analyze aspects of *all* scientific theories, once again only evolution was singled out for "critical analysis."

The ambiguities of this statement became clear when a board member, Tom McClain, was quoted online at *Christianity Today* saying that Standard 23 meant that students would be debating *details* of evolutionary theory ("What we're essentially saying here is evolution is a very strong theory, and students can learn from it by analyzing evidence as it is accumulated over time" [Olsen 2002]), and ID proponent Phillip Johnson claimed that the standard would promote the teaching of "evidence against evolution" (Staub 2002).

Disclaiming Evolution

Disclaimers often present evolution as "theory, not fact," using "theory" in the popular sense of "guess" or "hunch" rather than in the scientific sense of "explanation." Attempts to copy a 1994 Alabama State Board of Education disclaimer placed in textbooks have been made in other states and communities. (See Figure 9.1.)

From 1974 to 1984, the state of Texas had the following regulations regarding the purchasing of textbooks:

Figure 9.1
The Alabama Disclaimer

A MESSAGE FROM THE ALABAMA STATE BOARD OF EDUCATION

This textbook discusses evolution, a controversial theory some scientists present as a scientific explanation for the origin of living things, such as plants, animals and humans.

No one was present when life first appeared on earth. Therefore, any statement about life's origins should be considered as theory, not fact.

The word "evolution" may refer to many types of change. Evolution describes changes that occur within a species. (White moths, for example, may "evolve" into gray moths.) This process is microevolution, which can be observed and described as fact. Evolution may also refer to the change of one living thing to another, such as reptiles into birds. This process, called macroevolution, has never been observed and should be considered a theory. Evolution also refers to the unproven belief that random, undirected forces produced a world of living things.

There are many unanswered questions about the origin of life which are not mentioned in your textbook, including:

- Why did the major groups of animals suddenly appear in the fossil record (known as the "Cambrian Explosion")?
- Why have no new major groups of living things appeared in the fossil record for a long time?
- Why do major groups of plants and animals have no transitional forms in the fossil record?
- How did you and all living things come to possess such a complete and complex set of "Instructions" for building a living body?

<u>Study hard and keep an open mind</u>. Someday, you may contribute to the theories of how living things appeared on earth.

The Texas Disclaimer, 1974–1984

Textbooks that treat the theory of evolution shall identify it as only one of several explanations of the origins of humankind and avoid limiting young people in their search for meanings of their human existence.

(A) Textbooks presented for adoption which treat the subject of evolution substantively in explaining the historical origins of man shall be edited, if necessary, to clarify that the treatment is theoretical rather than factually verifiable. Furthermore, each textbook must carry a statement on an introductory page that any material on evolution included in the book is clearly presented as theory rather than verified.

(B) Textbooks presented for adoption which do not treat evolution substantively as an instructional topic, but make reference to evolution indirectly or by implication, must be modified, if necessary, to ensure that the reference is clearly to a theory and not to a verified fact. These books will not need to carry a statement on the introductory page.

(C) The presentation of the theory of evolution shall be done in a manner which is not detrimental to other theories of origin.

The state attorney general was requested to rule on the constitutionality of these regulations. Citing the Lemon decision, he judged them unconstitutional.

Texas Attorney General's Opinion JM-134, March 12, 1984

. . . The only aspect of "evolution" with which the rule is concerned is that which relates to "the historical origins of man." The rule requires a biology textbook, for example, to carry a disclaimer on its introductory page to the effect that "any material on evolution" included therein is to be regarded as theory rather than as factually verifiable. In the first place, such a disclaimer—which might make sense if applied to all scientific theories—is limited to one aspect—man's origins—of one theory—evolution—of one science—biology. In the context of the controversy between evolutionists and creationists which was before the board at the time of the rule's adoption both in 1974 and 1983, this singling out of one aspect of one theory of one science can be explained only as a response to pressure from creationists.

In the second place, the "theory of evolution," as it is commonly treated in biology texts, is a comprehensive explanation of the development of the various plant and animal species. Only a relatively minor portion is concerned with the "historical origins of man." The latter subject is the primary interest of creationists. Again, the inference is inescapable from the narrowness of the requirement that a concern for religious sensibilities, rather than a dedication to scientific truth, was the real motivation for the rules.

Finally, the rules require that a textbook identify the theory of evolution "as only one of several explanations" of human origins in order to "avoid limiting young people in their search for meanings of their human existence." (Emphasis added.) Such language is not conducive to an explanation that the purpose of the rule is to insure that impressionable minds will be able to distinguish between scientific theory and dogma. The "meaning of human existence" is not the stuff of science but rather, the province of philosophy and religion. By its injection into the rules language which is clearly outside the scope of science, the board has revealed the non-secular purpose of its rules.

Clearly, the board made an effort, as it has stated, to "insure neutrality in the treatment of subjects upon which beliefs and viewpoints differ dramatically." In our opinion, however, the board, in its desire not to offend any religious group, has injected religious considerations into an area which must be, at least in the public school context, strictly the province of science. . . .

If the board feels compelled to legislate in this area, it should, in order to avoid the constitutional prohibition, promulgate a rule which is of general application to all scientific inquiry, which does not single out for its requirement of a disclaimer a single theory of one scientific field, and which does not include language suggesting inquiries which lie totally outside the realm of science. The rules submitted, however, when considered in the context of the circumstances of their adoption, fail to evidence a secular purpose, and hence we believe a court would find that they contravene the first and fourteenth amendments to the United States Constitution.

Academic Freedom to Teach Creationism

After *McLean* and *Edwards*, court battles shifted from state-level regulations to court cases involving individual teachers or school districts seeking to teach some form of creationism or Neocreationism. These cases usually involved Establishment Clause concerns, but also Free Exercise and Free Speech (academic freedom) issues as well. Continuities with the past include the claim that the teaching of creationism or Neocreationism enhances students' rights to learn "all the material."

In response to student complaints, the New Lenox, Illinois, superintendent of education, Alex Martino, directed middle school teacher Ray Webster to cease teaching creationism in his classroom. Arguing academic freedom, Webster sued the district for his right to teach Creation Science. A student, Matthew Dunne, was a coplaintiff, arguing that he had a right to hear about Creation Science in school. Citing *Edwards*, the U.S. District Court ruled against both plaintiffs. The decision was appealed and upheld by the 7th Circuit Court of Appeals.

Webster v. New Lenox

Raymond Webster is a teacher for New Lenox and as such has certain responsibilities to teach within the framework of curriculum outlined by the District. . . . Webster's rights as a teacher to present certain material within his social studies curriculum is not absolute. . . .

If a teacher in a public school uses religion and teaches religious beliefs or espouses theories clearly based on religious underpinnings, the principles of the separation of church and state are violated as clearly as if a statute ordered the teacher to teach religious theories such as the statutes in *Edwards* did. The school district has the responsibility of ensuring that the Establishment Clause is not violated. Therefore, New Lenox has the responsibility of monitoring the content of its teachers' curricula to ensure that the establishment clause is not violated.

Although Webster denies any improper religious teaching, the question before this court is whether Webster has a first amendment right to teach creation science. As previously discussed, the term "creation science" presupposes the existence of a creator and is impermissible religious advocacy that would violate the first amendment. Webster has not been prohibited from teaching any nonevolutionary theories or from teaching anything regarding the historical relationship between church and state. Martino's letter of October 13, 1987, makes it clear that the religious advocacy of Webster's teaching is prohibited and nothing else. Since no other constraints were placed on Webster's teaching, he has no basis for his complaint and it must fail.

Plaintiff Dunne's claims, if not moot, are without merit. Dunne has not been denied the right to hear about or discuss any information or theory including information as to creation science. He is merely limited to receiving information as to creation science to those locations and settings where dissemination does not violate the first amendment. Dunne's desires to obtain this information in schools are outweighed by defendants' compelling interest in avoiding Establishment Clause violations and in protecting the first amendment rights of other students. Dunne simply fails to state a claim for the violation of any first amendment or other rights and thus his claim must also fail. . . .

The relevant issue here is what Webster was prohibited from teaching. He was prohibited from teaching Creation Science. The U.S. Supreme Court has found that Creation Science is a religiously based theory and that the teaching of this theory in a public school violates the First Amendment. Prohibiting this teaching is thus constitutionally valid.

Since plaintiff Webster has no right to teach Creation Science and plaintiff Dunne has no right to receive information regarding Creation Science in his public school room, both plaintiffs' actions must fail.

[Internal references deleted.]

REFERENCES CITED

Cornelius, R. M., ed. 1996. *Selected Orations of William Jennings Bryan: The Cross of Gold Centennial Edition*. Dayton, TN: William Jennings Bryan College.

Hansen, John. 1997. *Operation T.E.A.C.H.E.S. 1997 Progress Report*. Williams Bay, WI: John Hansen.

Hansen, John. 1999. *Operation T.E.A.C.H.E.S. 1998 Progress Report*. Williams Bay, WI: John Hansen.

Hansen, John. 2000. *Operation T.E.A.C.H.E.S. 2000 Progress Report*. Williams Bay, WI: John Hansen.

Hardy, Dan. 2002. District Is Open to Theories About Life. *Philadelphia Inquirer*, December 23, p. B-1.

Larson, Edward J. 2003. *Trial and Error: The American Controversy over Creation and Evolution*, rev. ed. New York: Oxford University Press.

Olsen, Ted. 2003. *Weblog: Ohio Science Standards Don't Mandate Intelligent Design, but May Open Door* [web site]. Christianity Today [cited Aug. 3 2003]. Available from http://www.christianitytoday.com/ct/2002/140/22.0.html.

Rosen, Laurel. 2003. Darwin Faces a New Rival; a Roseville High School Parent Urges That "Intelligent Design" Also Be Taught in Biology. *Sacramento Bee*, June 22, pp. 1, 2.

Available from http://www.sac.bee.com/content/community_news/roseville/roseville/story 6901910p-7851550c.html.
Staub, Dick. 2002. *The Dick Staub Interview: Phillip Johnson*. December 3, 2002. Accessed August 3, 2003. Available from http://www.christianity today.com/ct/2002/147/22.0.html.
Trevas, Dan. 1996. Rep: Teach Evolution as a Theory, Not Fact. *Vindicator*, April 28, p. B5.

Cases Cited

Edwards v. Aguillard. 482 U.S. 578 (1987).
Epperson v. Arkansas. 393 U.S. 97 (1968).
Lemon v. Kurtzman. 403 U.S. 602, 612–613 (1971).
McLean v. Arkansas Board of Education. 529 F.Supp. 1255 (E.D. La. 1982).
Webster v. New Lenox School District. 917 F2d 1004 (7th Cir. 1994).

Selection excerpted from:

Webster v. New Lenox School District 917 F2d 1004 (7th Cir. 1994).

CHAPTER 10

Educational Issues

INTRODUCTION

The antievolution movement has long targeted the teaching of evolution in the public schools. Most young people who encounter evolution will do so in schools, so that is where antievolutionists seek to ameliorate what they perceive as evolution's negative influence. As discussed in chapters 5 and 6, antievolutionists originally tried to ban evolution from the curriculum, then to "balance" it with the teaching of creationism or Creation Science. In response to court decisions, antievolutionists currently avoid religious arguments in favor of arguing that evolution is bad science that should be "balanced" with the presentation of "alternative scientific theories" such as Intelligent Design. They also argue that the teaching of evolution should be balanced with "evidence against evolution."

In this battle over curricula, three themes recur. One is that teaching creationism (or Creation Science) and evolution is a matter of fairness or equity. Because so many people object to their children learning evolution, many believe that if the subject is taught, it at least should be "balanced" by creationism or Creation Science. A second theme is to promote the "fairness" approach as having positive pedagogical value. Here it is claimed that students learning and evaluating "all the evidence" benefit by sharpening their critical thinking skills. Finally, ever since the Scopes era, textbooks' coverage of evolution has been controversial, and the amount and quality of evolution in textbooks has fluctuated over time in response to creationist pressure. These three topics are the subject of this chapter.

"FAIRNESS"

Antievolutionists often exhort teachers to be "fair" and present "both views," meaning creationism and evolution. As discussed in chapter 3, there are far more than two views, and no science classroom can possibly devote sufficient time to give justice to even a fraction of the various religious explanations of the universe. Creation Science proponents, on the other hand, argue that theirs is a legitimate *scientific* view, not merely a religious perspective, and that out of a sense of fairness, both "evolution science" and "Creation Science" should be presented. "Let the children decide" has been the rallying cry of Young Earth Creationists for decades.

The fairness argument appeals to Americans—indeed, North Americans in general—because of cultural traditions of free speech, individuality, town meetings, democratic traditions, and the like. The argument has been made, however, that science is not a democratic institution, and that cultural standards of equal time do not apply to empirical explanations that have been analyzed and rejected. Opponents of the fairness argument do not question the cultural value of fairness, but rather its application to the science classroom.

In Support of "Fairness" in Teaching Evolution

Henry M. Morris is the founder of modern twentieth century "creation science." In this appendix to a model resolution originally intended for adoption by state legislatures to encourage the teaching of creation science with evolution, he lays out very clearly the perspective that creation science deserves to be taught along with evolution for reasons of science, pedagogy, and fairness.

Appendix E: Fairness of a Balanced-Treatment Approach

In view of the fact that evolution and creation are the only two possible concepts of origins, that evolution requires at least as much of a "religious" faith as does creation, and that creation fits all the "scientific" data at least as well as does evolution, it is clear that *both* should be taught in the schools and other public institutions of our country, and that this should be done on an equal-time, equal-emphasis basis, in so far as possible.

This is obviously the only equitable and fair approach to take, the only one consistent with traditional American principles of religious freedom, civil rights, freedom of information, scientific objectivity, academic freedom, and constitutionality. That American citizens, when given opportunity to express their opinions, fully support this idea has been proven conclusively in recent carefully conducted, scientifically organized community polls taken in two California school districts.

One of these was a semi-rural district, Del Norte County in northern California. Here, a poll of 1,326 homes revealed 89 per cent to favor including creation along with evolution in school curricula (*Acts and Facts*, Vol. 3, April 1974, p. 1). The other was a very cosmopolitan district in the San Jose–San Francisco metropolitan region, Cupertino, the largest elementary school district in the state. In this case, a poll of over 2,000 homes showed 84 per cent to favor including creation (*Acts and Facts*, Vol. 3, August 1974, p. 3). In both cases, the emphasis in the questionnaire was on *scientific* creationism, rather than its religious aspects.

There is little doubt that similar majorities would be obtained in most other school districts across the country, if people were informed on the issue and given opportunity to express their preferences.

Two final points should be stressed. There are really only two scientific models of origins—continuing evolution by natural processes or completed creation by supernatural processes. The latter need not be formulated in terms of Biblical references at all, and is not comparable to the various cosmogonic myths of different tribes and nations, all of which are merely special forms of evolutionism, rather than creationism, rejecting as they do the vital creationist concept of a personal transcendent Creator of all things in the beginning. It is not the Genesis story of creation that should be taught in the schools, of course, but only creationism as a scientific model.

Secondly, the idea of theistic evolution (that is, evolution as God's method of creation) is not in any way a satisfactory compromise between evolution and creation. It is merely an alternate form of evolutionism with no *scientific* distinction from that of naturalistic evolutionism, and is vulnerable to all the scientific, religious and legal objections outlined previously for evolutionism in general.

We conclude, therefore, that *both* creation and evolution should be taught—as scientific models only—in all books and classes where *either* is taught or implied. Administrators should assume the responsibility of providing adequate training and materials to enable their teachers to accomplish this goal.

Selection excerpted from:

Morris, Henry M. 1975. ICR. *Impact* 26 (August, p.4).

Opposing "Fairness" in Teaching Evolution

It's Only Fair to Teach Creationism

Teachers need to be able to counter the "fairness" argument, which is second only to the idea that evolution is antireligious as a motivator of pro-creationism and anti-evolution activities. . . . What are some answers when this question arises in a community or classroom?

1) *We determine curricula based on the best scholarship, not because of political pressure.* Both scientists and teachers have rejected "creation science," "intelligent design theory," and other forms of creationism as science. There are many statements from scientific and educational organizations stating that religiously-based ideas should not be taught as science; only evolution should be taught as science (these are also on the World Wide Web at www.ncseweb.org/voices.htm). Creation science and other forms of creationism should not be taught because they have been evaluated by scientists and educators and found wanting. . . . Just because a pressure group wants to change the curriculum is no reason to abandon scholarly standards.

2) *Science is not a democratic process; nature doesn't work according to our wishes!* It would be a wonderful solution to the problem of universal cheap energy if perpetual motion were a reality, but no matter how many people want this to be so, the laws of

physics do not allow it, and we do not teach students perpetual motion as a scientific principle. Even if many individuals in a community favor a special creationist view that the universe came into being in its present form 10,000 years ago, scientific evidence is very much against this, and students should not be taught that special creationism is a viable scientific doctrine.

3) *It is not "fair" to mislead students by presenting scientifically uncontroversial issues as controversial*. Evolution is taught matter-of-factly at any reputable university in this country, including Brigham Young (Mormon), Baylor (Baptist), and Notre Dame (Catholic). It is a controversial subject only to members of the public. It would be "fair" to discuss the creation and evolution controversy as a politically- or socially-controversial issue, but it shortchanges students to teach them that there is any scientific controversy over the concept that the universe has changed through time, and that living things have descended with modification from earlier ancestors. "Equal time for creationism" policies do just that, handicapping students for future study in college, and decreasing their scientific literacy (pp. 73–74).

[Citations omitted.]

Selection excerpted from:

Scott, Eugenie C. 1999. "But I Don't Believe in Evolution!" The Science Teacher's Dilemma. *Journal of Religion and Education* 26 (2): 67–75.

THE PEDAGOGICAL VALUE OF TEACHING "ALL THE EVIDENCE"

Most North American science standards on the state or provincial level stress the importance of critical thinking. One approach to critical thinking often used by teachers is to have students research and debate a controversial issue. But is the creation and evolution issue an appropriate one for critical thinking exercises? The first two readings take the position that it is.

We had intended to include an editorial written by Intelligent Design proponent Stephen C. Meyer during a controversy over the Ohio state science education standards (*Cincinnati Enquirer*, March 30, 2002), but permission was denied. Students are encouraged to read the article online at http://www.discovery.org/viewDB/index.php3?program=CRSC& command=view&id=1134.

In this article, Meyer begins by stating a general principle of "teaching the controversy": when two groups of experts disagree about a controversial topic relevant to a subject taught in the public schools, teachers should explain the controversy, and the arguments on both sides of it, to their students. Speaking to the Ohio Board of Education while the state was considering new state science standards, Meyer recommended teaching the controversy as a way of resolving disputes about how to teach about evolution and whether or not to teach Intelligent Design alongside what he calls "Darwinism."

In particular, he suggested (1) that teachers ought not to be required to teach Intelligent Design yet, (2) that teachers ought to teach the scientific controversy about Darwinian evolution, and (3) that teachers ought to be allowed, but not required, to teach Intelligent Design. He cited five considerations in favor of his recommendation. First, that there is a genuine scientific controversy about Darwinian evolution, as evidenced by a bibliography of scientific articles prepared by the Discovery Institute and submitted to the Ohio Board of Education and by the Discovery Institute's "A Scientific Dissent from Darwinism," a statement on "Darwinism" signed by 100 scientists. Second, the Supreme Court's decision in *Edwards v. Aguillard* noted that "teaching a variety of scientific theories about origins" was legal. Meyer recommended teaching critiques of "Darwinism" and "discussion of competing theories." Third, that it is endorsed by the Santorum language in the conference report to the No Child Left Behind Act (see chapter 9). Fourth, that it is overwhelmingly approved of by voters, according to a national Zogby poll. Fifth, that teaching the controversy is educationally sound: it sparks interest among students and motivates them to learn. Here Meyer quoted from Darwin: "A fair result can be obtained only by fully stating and balancing the facts and arguments on both sides of each question." He concluded by saying that the issue in Ohio is not about Intelligent Design but rather about whether both sides of the scientific controversy over Darwinism will be taught.

We had intended to reprint a portion of an essay from Intelligent Design proponent David DeWolf (David K. DeWolf, *Teaching the Origins Controversy: A Guide for the Perplexed.* Discovery Institute Special Report. Accessed November 20, 2002. Available from http://www.discovery.org/viewDB index.php3?program=CRSC&command=view&id=48), but we were denied permission. Students are encouraged to read the original themselves.

The passage summarized here begins with the same point from the Supreme Court's decision in *Edwards v. Aguillard* emphasized by Meyer: that "teaching a variety of scientific theories . . . might be validly done with the clear secular intent of enhancing the effectiveness of science education." DeWolf suggests that teaching the origins controversy—or, as he says, "making a full, rather than restricted[,] presentation of Darwinism"—would do so. Darwinism, he contends, is typically taught in a way that fails to stimulate students' interest. It ought not to be taught as uncontroversial and as accepted by all reputable scientists, as the National Academy of Sciences suggests. Rather, he suggests, it ought to be presented as a competitor in the scientific arena, vying with other plausible scientific theories for acceptance. Students would thereby be motivated to scrutinize the arguments for and against these theories carefully. Moreover, DeWolf suggests, teaching the origins controversy provides teachers with the opportunity to encourage their students to examine both sides of a controversial issue; even if teachers have strong views on the issue, they can set a good example by forthrightly addressing the view they reject. Learning is facilitated best by active dialogue rather than passive acceptance.

Opposing the Pedagogical Value of Teaching "All the Evidence"

The following selection disputes the idea that teaching "two models" or evolution and antievolutionism is good pedagogy.

Evolution: What's Wrong with "Teach the Controversy?"

. . . In a piece entitled "Teach the Controversy," Stephen C. Meyer of the Discovery Institute's Center for Science and Culture, the institutional home of the "intelligent design" variety of antievolutionism, writes, "good pedagogy commends this approach. Teaching the controversy about Darwinism as it exists in the scientific community will engage student interest. It will motivate students to learn more about the biological evidence as they see why it matters to a big question." The thought does not originate with the "intelligent design" movement, however. The Institute for Creation Research (ICR), the oldest major antievolutionist organization in the USA, recommends that students and teachers be "encouraged to discuss the scientific information that *supports* and *questions* evolution and its underlying assumptions, in order to promote the development of critical thinking skills" (emphasis in original). The intent is not to have students investigate controversies about patterns and processes within evolutionary theory, but to debate whether evolution occurred.

Presenting all sides of a controversial issue appeals to popular values of fairness, openness and equality of opportunity. It thus plays well with the public. But it is important to examine any such appeal carefully, because it is easy to abuse the public's willingness to be swayed by such a call. . . . How is a teacher to decide which controversies are pedagogically valuable?

Criteria for Determining Which Controversies to Teach

We suggest the following five criteria for whether a controversy is appropriate to teach in a public school science class:

The controversy ought to be of interest to students.

There is, for example, a raging scientific controversy over whether maximum likelihood or parsimony ought to dominate in phylogenetic interpretation. But we suspect that few students will be fascinated by the controversy, however dear it might be to the readers of *TREE*!

The controversy ought to be primarily scientific, rather than primarily moral, social or religious.

The controversy over stem cell research, for example, is not about whether embryos can be manipulated to produce stem cells, but about whether it is morally permissible to do so. Questions about the morality of such research are of course important, but they are not suitable for a science class. Controversies that are primarily religious in nature are especially unsuitable for classes in public schools in the USA, due to the Establishment Clause of the First Amendment to the Constitution, which prohibits the government from sponsoring religious advocacy.

The resources for each side of the controversy ought to be comparable in availability.

It is difficult to teach the controversy if there is hardly anyone to make the case for one side of it. A teacher who decided to teach the controversy about geocentrism, for example, would find it difficult to locate resources for the geocentric side. (It would, however, be appropriate to teach about the 17th-century controversy as an historical digression.)

The resources for each side of the controversy ought to be comparable in quality.

If the arguments for one side of a controversy are generally poor, then students are not likely to profit by studying it. The scientific consensus that AIDS is caused by a virus, for example, is so strong that there is little point to presenting opposing views.

The controversy ought to be understandable by the students.

Most of the fascinating controversies over the role of epigenetic factors in development, for example, require a great deal more developmental, morphological and genetic training than a high school student can be expected to master in the time available.

Using these criteria, is the antievolutionists' controversy about evolution one that is worth teaching? We think not. It does satisfy criterion 1: it is probably of interest to students. It also satisfies criterion 3, thanks both to the wide availability of creationist material on the internet and to the advent of "intelligent design," which enjoys a degree of publicity in relatively mainstream venues, although conspicuously not in the peer-reviewed scientific literature. However, the controversy about evolution fails significantly to satisfy the other three criteria.

First, the controversy is not primarily scientific (criterion 2). In spite of their frequent claims to be concerned with the science, for young-earth creationists (such as the ICR) and the "intelligent design" movement alike, the science is essentially a smokescreen for nonscientific concerns. For the ICR, the problem with evolution is its incompatibility with a literal reading of the Bible; for the "intelligent design" movement's guru Phillip Johnson, "[t]his isn't really, and never has been, a debate about science. It's about religion and philosophy." Moreover, as far as students are concerned, the controversy about evolution is essentially religious. Are they going to be able to restrict their concerns solely to the science? Even many college students have difficulty studying religion objectively; at the pre-college level, the problem is worse. And are science teachers willing and able to respond properly to their concerns, without appearing to attack religious beliefs?

. . . Second, and correspondingly, the scientific quality of the antievolutionist resources is exceedingly poor (criterion 4). The positive claims of young-earth creationism—that the universe and the Earth were created ~10,000 years ago, that the Earth was inundated by Noah's Flood, and that all living things were created by God to reproduce "after their kind," thus setting limits on evolution—are unanimously rejected by the scientific community. Its negative claims (the "evidence against evolution") typically involve either misinterpretation of the scientific literature or arguments from ignorance. The "intelligent design" variety of antievolutionism is strategically noncommittal, limiting its positive program to the claim that it is possible to identify certain natural phenomena as the products of intelligent design; its proponents disagree about the age of the Earth, common ancestry, and a host of other important scientific issues. Its negative claims are already in the repertoire of young-earth creationism, so the same objection applies.

Finally, students are unlikely to be able to understand both sides of the controversy (criterion 5). The evidence for evolution is easy to understand, at least on a basic level. But the antievolutionist critique of evolution ranges freely and opportunistically through the scientific literature, from astronomy to zymurgy, frequently misrepresenting it in the process. Faced with the ICR's tendentious and eclectic list of "questions that could be used to critically examine and evaluate evolutionary theory" or the "Suggested Warning Labels for Biology Textbooks" produced by "intelligent design" proponent Jonathan Wells, even a working research scientist would have a difficult time sorting through the quagmire of misleading and mistaken claims. It is unreasonable to expect teachers, much less their students, to do so . . . (499–501).

[Citations deleted.]

Selection excerpted from:

Scott, Eugenie C., and Glenn Branch. 2003. Evolution: What's Wrong with "Teaching the Controversy?" *Trends in Ecology and Evolution* 8 (10): 499–502.

• •

TEXTBOOK BATTLES

For better or for worse, textbooks tend to comprise the de facto curriculum in science classes. When textbooks include evolution, antievolutionists are motivated to modify them to either delete or water down evolution (as during the post-Scopes era), or "balance" it with "alternatives to evolution" (including creationism) or, most recently, to present evolution and the "scientific weaknesses of evolution." This first selection traces the history of evolution in textbooks.

Effects of the Scopes Trial

. . . The scientific community in the 1920's responded forcefully to the overt attack in the Scopes case. But it failed to follow through. As a result, the teaching of evolution in the high schools—as judged by the content of the average high school biology textbooks—*declined* after the Scopes trial (p. 832).

. . . The impact of the Scopes trial on high school biology textbooks was enormous. It is easy to identify a text published in the decade following 1925. Merely look up the word "evolution" in the index or the glossary; you almost certainly will not find it (p. 833).

. . . We do not wish to maintain that the omission of the word "evolution" from the index of a book automatically invalidates the book's treatment of the subject. However, widely used biology textbooks of the 1930's did not treat evolution very well in the text, either. The religious quotations which appear in some of these books, together with the near-disappearance of the theory of evolution and of Darwin's role in establishing it, demonstrate the impact of fundamentalist pressure in general, and the Scopes trial in particular, on the textbook industry (pp. 835–836).

. . . Publishing high school textbooks is a lucrative business. And the authors and publishers of biology textbooks have to pay attention to their market. Textbook adoption practices vary; some states approve texts for the entire state, while others allow local option. Unfortunately for the market prospects of an evolutionary textbook, most of the states which have at various times practiced statewide textbook adoption are in the south, and no eastern states are included. . . .

Publishers and authors feared that a good treatment of evolution meant the loss of the southern market—a fear which seems to have been justified (p. 835).

. . . The evolutionists of the 1920's believed they had won a great victory in the Scopes trial. But as far as teaching biology in the high schools was concerned, they had not won; they had lost. Not only did they lose, but they did not even know they had lost. A major reason was that they were unable to understand—sympathetically or otherwise—the strength of the opponents of evolution. It is worth one's while to inquire into what motivates large numbers of people to oppose evolution. Whether one agrees or disagrees with their views, the people and their concerns deserve sympathy and respect. And understanding the opposition to evolution is essential if one is to take any kind of effective action (p. 836).

. . . Readers may choose their own villain in the story we have told. Like us, some will find the greatest culpability in the scientific community itself, for the large-scale failure to pay attention to the teaching of science in the high schools. Others will blame the textbook authors and publishers for pursuing sales rather than quality. Some will attach blame to the politicians who exploited antievolution sentiment to get into, or remain, in office. Others will blame the conservative Protestant clergy. Some may blame the whole educational system for failing to teach Americans how to evaluate evidence. And many will blame the evolutionists for bringing the matter up in the first place. But whatever the lesson one wishes to draw from the history of biology textbooks since the Scopes trial, we think the story itself is worth knowing. That the textbooks could have downgraded their treatment of evolution with almost nobody noticing is the greatest tragedy of all (pp. 836–837).

[Citations omitted.]

Selection excerpted from:

Grabiner, Judith V., and Peter D. Miller. 1974. Effects of the Scopes Trial. *Science* 185 (4154): 832–837.

Criticisms of the Coverage of Evolution in Textbooks

Jonathan Wells's book *Icons of Evolution* criticizes textbooks' coverage of evolution. He claims that a series of "icons" or common illustrations of evolution or mechanisms of evolution are misleadingly presented in textbooks, even to the extent of presenting fraudulent data. One example is the familiar peppered moth story of natural selection. We had intended to excerpt an article by Wells that summarized his views about the peppered moth, but were denied permission to do so. Readers are encouraged to read the article ("Survival of the Fakest," *The American Spectator*, December

2000/January 2001; online at http://www.discovery.org/articleFiles/PDFs/survivalOfTheFakest.pdf) or the relevant chapter in *Icons of Evolution*.

The section of Wells's article titled "Nothing a Little Glue Can't Fix: The Peppered Moths" begins by observing that Darwin's *On the Origin of Species* lacked a concrete example of natural selection in action, and then turns to a discussion of Bernard Kettlewell's research in the 1950s on peppered moths. During the nineteenth century, the British population of these moths changed from being mostly light-colored to being mostly dark-colored. The cause of the change was thought to be natural selection: due to industrial pollution, the tree trunks on which the peppered moths rest became darker, so light-colored moths were more susceptible to predation by birds. Kettlewell tested this hypothesis by releasing light-colored and dark-colored moths in polluted and unpolluted woods; the moths that were not appropriately camouflaged were indeed more susceptible to predation. Kettlewell's study was widely heralded as a classic example of natural selection and is widely cited in biology textbooks, which typically include photographs of moths resting on tree trunks.

Further research conducted in the 1980s revealed flaws in Kettlewell's study, however: although he released the moths by day onto tree trunks, they normally fly at night and rest under upper tree branches during the day. Consequently, Wells states, many biologists now reject Kettlewell's study, and some even doubt the hypothesis that natural selection caused the change in moth coloration. The textbook photographs were staged; in some cases, the moths were glued to the trees. Although the flaws in Kettlewell's study were discovered in the 1980s, textbooks continue to recount the study uncritically and to reproduce the staged photographs. Wells quotes a textbook author as justifying the continued use of Kettlewell's study by saying, "We want to get across the idea of selective adaptation. Later on, they can look at the work critically." He then observes that Jerry Coyne, a professor of biology at the University of Chicago, learned the truth about the peppered moths only in 1998, showing that the "icons of evolution" (of which the peppered moth example is one) are so deceptive that even experts such as Coyne are apt to be misled.

Responses to the Critics

Evolutionary biologist Jerry Coyne responded to Wells's reporting of his ideas in a letter to the editor written during a Kansas controversy over the state science education standards.

Coyne on Wells and Moths

My only problem with the peppered moth story is that I am not certain whether scientists have identified the precise agent causing the natural selection and evolutionary change. It may well be bird predators, but the experiments leave room for doubt. Creationists such as Jonathan Wells claim that my criticism of these experiments casts strong doubt on Darwinism. But this characterization is false. All of us in the peppered moth debate agree that the moth story is a sound example of evolution produced by natural selection. My call for additional research on the moths has been wrongly characterized by creationists as revealing some fatal flaw in the theory of evolution.

Selection excerpted from:

Coyne, Jerry. 2000. Letter to the editor. *Pratt* (Kansas) *Tribune*, December 6.

In response to Jonathan Wells's book Icons of Evolution, *Alan D. Gishlick prepared a detailed Internet-based critique of each of the "icons." The following is an excerpt from the chapter analyzing Wells's treatment of the peppered moth.*

How Many Moths Can Dance on the Trunk of a Tree? Distraction by Irrelevant Data

[Wells's] discussion centers on three points where he believes textbooks are in error, alleging that (1) the daytime resting places of peppered moths invalidates Kettlewell's experimental results; (2) the photos of the moths are "staged"; and (3) the recovery patterns of populations dominated by light moths after the levels of pollution were reduced do not fit the "model," although he is unclear as to what the "model" is. All three of these objections are spurious. They are distractions from the general accuracy of the story and its value in showing the effects of natural selection on genetic variability in natural populations.

First, Wells argues that the story is seriously flawed because "peppered moths in the wild don't even rest on tree trunks." He repeats this point throughout the chapter. However, it is both false and irrelevant, and only serves as a distraction to lead the reader away from the actual story of the moths. Contrary to Wells's assertions, data given by Majerus indicate that the moths do indeed rest on the trunks of trees 25% of the time. The rest of the time moths rest in branches (25%) or at branch-trunk junctions (50%). . . . Moths are found all over trees, which is not a surprise and it is mentioned in the references that Wells cites.

To clear up any confusion, no researcher doubts that the peppered moth rests in trees, which means that the resting substrate is bark. Entire trees are stained by pollution . . . and so the colors of the moths are relevant no matter where on the tree they rest—trunks, trunk–branch junctions, branches, twigs, and even the leaves. Wells's argument implies that predatory birds can only see moths that are on exposed trunks. By making this argument, however, Wells shows an apparent ignorance of the ecology of birds and woodland ecosystems. If you walk into any forest, you can see that the birds fly from tree to tree, branch to branch, and hunt at all levels of the forest. Woodland species of birds that prey on moths and other insects live and hunt in the canopy (the leafy part of the trees). These birds are not hunting from outside, soaring above the trees like hawks, as Wells's argument would require.

. . . The purpose of Wells's distraction is to put the actual experiments into question and make it sound as if the textbook authors are either mistaken, or intentionally trying to fool students. The insinuation is that because Kettlewell released the moths during the day, they did not find "normal" resting places. Whether or not this is so, the release and capture experiments took place over a number of days, so the moths were able to take up positions of their choosing, even if the first day was not perfectly "natural." Kettlewell's experiments were not perfect—few field experiments are—and they may have magnified the degree of selection, but all serious researchers in the field agree that they were certainly not so flawed as to invalidate his conclusion.

In his second objection, Wells ties the Kettlewell experiments to textbooks by constantly repeating the statement that the illustrative photos were "staged"; the important issue here is not how the photos were made, but rather their intent. Wells implies that the photos purport to show a "lifelike" condition to prove that moths rest on trunks. This is not the case. The photos are meant to demonstrate the visibility of the different forms of the moth on polluted and unpolluted trees. It is absurd to expect a photographer to just sit around and wait until two differently colored moths happen to alight side by side. Further, how the photos were produced does not change the actual data. Birds eat moths and they eat the ones that they see more easily first. The textbook photos never claim to depict a real-life situation, and it is improper to imply otherwise.

The third criticism, and the only scientific one that Wells levels, deals with the recovery of the light form of the moth following the institution of pollution control laws. The main thrust of his argument is that because the recovery of light-colored lichens does not correlate with the recovery of the light form of the moths, the entire story is incorrect. Wells exploits the fact that the original researchers thought that the camouflage of the light moths depended on the presence of lichen. However, the light forms recovered before the lichens did; therefore, Wells concludes, natural selection has nothing to do with the story. Although it is true that the moths are well-camouflaged against lichens, and lichens are destroyed by pollution, nevertheless the camouflage of the moths ultimately depends upon the color of the trees, which reflect[s] the amount of soot staining the trees. Although lichens play a role in camouflage, they are not necessary. This is what happened: pollution was reduced, the trees got lighter, then the moths got lighter. Further, in all areas, the light moths have recovered, as predicted by the hypothesis. This is clearly stated in the literature, but it does not fit Wells's story, and he just ignores it.

[References omitted.]

Selection excerpted from:

Gishlick, Alan D. 2003. *Icons of Evolution? Why Much of What Jonathan Wells Writes About Evolution Is Wrong.* Icon no. 6: The Peppered Moth. National Center for Science Education, Inc. Accessed September 1, 2003. Available from http://www.ncseweb.org/icons/icon6tol.html.

Finally, a comment from the aforementioned Dr. Gishlick about the purpose of textbooks and the use of familiar examples.

Textbook "Icons": Why Do We Have Them?

Paradigms and all their components are not necessarily simple. To understand the depth of any scientific field fully requires many years of study. It is the goal of elementary and secondary education to give students a basic understanding of the "world as we know it," which includes teaching students the paradigms of a number of fields of science. In order to do this, teaching examples must be found. It is this need to find simple, easy-to-explain, dynamic, and visual examples to introduce a complex topic to students that has led to the common use of a few examples—the "icons." Yet, with our knowledge of the natural world expanding at near-exponential rates, the volume of new information facing a textbook writer is daunting. The aim of a textbook is not necessarily to report the "state of the art" as much as it is to offer an introduction to the basic principles and ideas of a certain field. Therefore, it should not be surprising that introductory textbooks are frequently simplified and may be somewhat out-of-date. In *Icons of Evolution*, however, Wells makes an even stronger accusation. Wells says: "Students and the public are being systematically misinformed about the evidence for evolution" through biology textbooks. This is a serious charge; to support it demands the highest level of scholarship on the part of the author.

Does Wells display this level of scholarship? Is Wells right? Are the "icons" out-of-date and in need of removal? And more importantly, is there something wrong with the theory of evolution?

In the following sections, each textbook "icon" is reexamined in light of Wells's criticism. The textbooks covered by Wells are examined as well, along with the grading criteria (given in the appendix of *Icons* and on the Discovery Institute's website) that he used to assess their accuracy. What was found is that although the textbooks could always benefit from improvement, they do not mislead, much less "systematically misinform," students about the theory of biological evolution or the evidence for it. Further, the grading criteria Wells applied are vague and at times appear to have been manipulated to give poor grades. Many of the grades given are not in agreement with the stated criteria or an accurate reading of the evaluated text. Beyond that, *Icons of Evolution* offers little in the way of suggestions for improvement of, or changes in, the standard biology curriculum. When Wells says that textbooks are in need of correction, he apparently means the removal of the subject of evolution entirely or the teaching of "evidence against" evolution, rather than the fixing of some minor errors in the presentation of the putative "icons." This makes *Icons of Evolution* useful at most for those with a certain political and religious agenda, but of little value to educators.

Selection excerpted from:

Gishlick, Alan D. 2003. *Icons of Evolution? Why Much of What Jonathan Wells Writes About Evolution Is Wrong. Introduction.* National Center for Science Education, Inc. Accessed September 1, 2003. Available from http://www.ncseweb.org/icons/index.htm.

CHAPTER 11

Religious Issues

INTRODUCTION

The creationism/evolution controversy exists in the United States because evolution is incompatible with certain religious views, and has profound implications for others. But as discussed in chapter 3, there are many religious perspectives, and many of them—even if we limit ourselves only to Christianity among the many religions practiced in the United States—embrace evolution. In this chapter, we will consider ways of relating science and religion, at literalist and nonliteralist approaches to creation, and at what is perhaps the overriding issue in the creationism/evolution controversy, the question of whether there is an existential loss of purpose or meaning to life if evolution occurred.

MODELS OF SCIENCE/RELIGION INTERACTION

Ian G. Barbour is the dean of science and religion studies. A pioneer in the field, he began writing on the topic in the 1950s, and continues to contribute ideas and stimulate thought in the field today. His model of science and religion interaction is widely used. Others have devised more complicated taxonomies, but all appear to include at least the relationships first outlined by Barbour.

When Science Meets Religion

In 1990, in the first chapter of *Religion in an Age of Science*, I proposed a fourfold typology as an aid to sorting out the great variety of ways people have related science and religion. I kept the same classifications with only minor modifications in the revised and enlarged edition of the book in 1997. In the present volume this typology is used as the organizing structure for every chapter.

1. *Conflict.* Biblical literalists believe that the theory of evolution conflicts with religious faith. Atheistic scientists claim that scientific evidence for evolution is incompatible with any form of theism. The two groups agree in asserting that a person cannot believe in both God and evolution, though they disagree as to which they will accept. For both of them, science and religion are enemies. These two opposing groups get most attention from the media, since a conflict makes a more exciting news story than the distinctions made by persons between these two extremes who accept both evolution and some form of theism.

2. *Independence.* An alternative view holds that science and religion are strangers who can coexist as long as they keep a safe distance from each other. According to this view, there should be no conflict because science and religion refer to differing domains of life or aspects of reality. Moreover, scientific and religious assertions are two kinds of language that do not compete because they serve contrasting questions. Science asks how things work and deals with objective facts; religion deals with values and ultimate meaning. Another version of the Independence thesis claims that the two kinds of inquiry offer complementary perspectives on the world that are not mutually exclusive. Conflict arises only when people ignore these distinctions—that is, when religious people make scientific claims, or when scientists go beyond their area of expertise to promote naturalistic philosophies. We can accept both science and religion if we keep them in separate watertight compartments of our lives. Compartmentalization avoids conflict, but at the price of preventing any constructive interaction.

3. *Dialogue.* One form of dialogue is a comparison of the methods of the two fields, which may show similarities even when the differences are acknowledged. For example, conceptual models and analogies are used to imagine what cannot be directly observed (God or a subatomic particle, let us say). Alternatively, dialogue may arise when science raises at its boundaries limit questions that it cannot itself answer (for example, Why is the universe orderly and intelligible?). A third form of dialogue occurs when concepts from science are used as analogies for talking about God's relation to the world. The communication of information is an important concept in many sciences; the pattern of unrepeatable events in cosmic history might be interpreted as including a communication of information from God. Or God can be conceived to be the determiner of the indeterminacies left open by quantum physics, without any violation of the laws of physics. Both scientists and theologians are engaged as dialogue partners in critical reflection on such topics, while respecting the integrity of each other's fields.

4. *Integration.* A more systematic and extensive kind of partnership between science and religion occurs among those who seek a closer integration of the two disciplines. The long tradition of natural theology has sought in nature a proof (or at least suggestive evidence) of the existence of God. Recently astronomers have argued that the physical constants in the early universe appear to be fine-tuned as if by design. If the expansion rate one second after the Big Bang had been ever so slightly smaller, the universe would have collapsed before the chemical elements needed for life could have formed; if the expansion rate had been even slightly higher, the evolution of life could not have occurred. Other authors start

from a particular religious tradition and argue that some of its beliefs (ideas of divine omnipotence or original sin, for instance) should be reformulated in the light of science. Such an approach I call a *theology of nature* (within a religious tradition) rather than a *natural theology* (arguing from science alone). Alternatively, a philosophical system such as process philosophy can be used to interpret scientific and religious thought within a common conceptual framework. It will be evident that my own sympathies lie with Dialogue and Integration (especially a theology of nature and a cautious use of process philosophy), but I hope that I have accurately described all four positions.

In 1995 John Haught offered a slightly different typology—one that may be easier to remember because all the terms start with the same letter. His first two categories, Conflict and Contrast, are identical with those in my scheme. His third category, Contact, combines most of the themes in what I have called Dialogue and Integration. He introduces a fourth heading, Confirmation, by which he means not the confirmation of particular theological doctrines (as one might assume) but rather the vindication by science of background assumptions originally derived from theology—for example, belief in the rationality and intelligibility of the world, which I treat as a form of Dialogue (pp. 1–4).

Selection excerpted from:

Barbour, Ian G. 2000. *When Science Meets Religion*. San Francisco: HarperSan Francisco. Reprinted by permission.

• •

LITERALIST VS. NONLITERALIST APPROACHES TO CREATION

Biblical literalists reject evolution because it is incompatible with an interpretation of Genesis that sees creation occurring by the direct hand of God, over a short (six-day) period of time. The literalist perspective is shared by a large minority of American Christians, though it is not the perspective of Catholics or mainstream Protestants. (A note to the reader: When references are made to verses in the Bible, the first number refers to the chapter, and the second to the verse. So Genesis 1:4, for example, refers to chapter 1 of Genesis, verse 4.)

Literalist Perspective on Evolution

As illustrated by minister/theologian John MacArthur, the major concern of biblical literalists is whether, if Genesis is not taken literally, the rest of the Bible can be taken literally (including miracles and the resurrection of Jesus).

Creation: Believe It or Not

The starting point for Christianity is not Matthew 1:1, but Genesis 1:1. Tamper with the Book of Genesis and you undermine the very foundation of Christianity. You cannot treat Genesis I as a fable or a mere poetic saga without severe implications for the rest of Scripture. The creation account is where God starts His account of history. It is impossible to alter the beginning without impacting the rest of the story—not to mention the ending. If Genesis I is not accurate, then there's no way to be certain that the rest of Scripture tells the truth. If the starting point is wrong, then the Bible itself is built on a foundation of falsehood.

In other words, if you reject the creation account in Genesis, you have no basis for believing the Bible at all. If you doubt or explain away the Bible's account of the six days of creation, where do you put the reins on your skepticism? Do you start with Genesis 3, which explains the origin of sin, and believe everything from chapter 3 on? Or maybe you don't sign on until sometime after chapter 6, because the Flood is invariably questioned by scientists, too. Or perhaps you find the Tower of Babel too hard to reconcile with the linguists' theories about how languages originated and evolved. So maybe you start taking the Bible as literal history beginning with the life of Abraham. But when you get to Moses' plagues against Egypt, will you deny those, too? What about the miracles of the New Testament? Is there any reason to regard *any* of the supernatural elements of biblical history as anything other than poetic symbolism?

After all, the notion that the universe is billions of years old is based on naturalistic presuppositions that (if held consistently) would rule out all miracles. If we're worried about appearing "unscientific" in the eyes of naturalists, we're going to have to reject a lot more than Genesis 1–3.

Once rationalism sets in and you start adapting the Word of God to fit scientific theories based on naturalistic beliefs, there is no end to the process. If you have qualms about the historicity of the creation account, you are on the road to utter Sadduceeism—skepticism and outright unbelief about *all* the supernatural elements of Scripture. Why should we doubt the literal sense of Genesis 1–3 unless we are also prepared to deny that Elisha made an axe-head float or that Peter walked on water or that Jesus raised Lazarus from the dead? And what about the greatest miracle of all—the resurrection of Christ? If we're going to shape Scripture to fit the beliefs of naturalistic scientists, why stop at all? Why is one miracle any more difficult to accept than another?

And what are we going to believe about the end of history as it is foretold in Scripture? All of redemptive history ends, according to 2 Peter 3:10–12, when the Lord uncreates the universe. The elements melt with fervent heat, and everything that exists in the material realm will be dissolved at the atomic level, in some sort of unprecedented and unimaginable nuclear meltdown. Moreover, according to Revelation 21:1–5, God will immediately create a new heaven and a new earth (see Isaiah 65:17). Do we really believe He can do that, or will it take another umpteen billion years of evolutionary processes to get the new heaven and the new earth in working order? If we really believe He can destroy *this* universe in a split second and immediately create a whole new one, what's the problem with believing the Genesis account of a six-day creation in the first place? If He can do it at the end of the age, why is it so hard to believe the biblical account of what happened in the beginning?

So the question of whether we interpret the Creation account as fact or fiction has huge implications for every aspect of our faith. These implications will become even more clear as we work our way through the text to the biblical account of Adam's fall. But the place to hold the line firmly is here at Genesis 1:1.

And that is no oversimplification. Frankly, believing in a supernatural, creative God who made everything is the only possible rational explanation for the universe and for life itself. It is also the only basis for believing we have any purpose or destiny (pp. 44–45).

Selection excerpted from:

MacArthur, John. 2001. *The Battle for the Beginning: The Bible on Creation and the Fall of Adam.* Nashville, TN: W. Publishing Group.

Nonliteralist Perspective on Evolution

The theologian Conrad Hyers provides a quite different perspective on the meaning of Genesis, derived through looking at the passages in historical and cultural context.

Comparing Biblical and Scientific Maps of Origin

. . . Now, the biblical accounts of creation in Genesis are different ways of mapping origins than those to which we who have been schooled in science are accustomed. In fact, even the two accounts of creation in Genesis 1 and 2 (the six-day account and the Adam-and-Eve account) have significant differences, reflecting the significant differences between the two cultural traditions in ancient Israel, the agricultural/urban and the shepherd/nomadic. Genesis 1 is a mapping of creation using the imagery, terminology, and perspectives of agricultural/urban Israel; and Genesis 2, of pastoral/nomadic Israel.

The Two Creation Accounts

We immediately recognize this difference in biblical language and usage elsewhere, as among the prophets and psalmists, to depict God's relationship to the world and to humanity. Isaiah, for example, goes back and forth between these two sets of imagery, as in Isaiah 40. On the one hand, Isaiah draws upon agricultural/urban imagery, as he speaks of God surveying the universe ("who has measured the waters in the hollow of his hand, and marked off the heavens with a span," v. 12), or God laying a foundation ("Have you not understood from the foundations of the earth," v. 21). But Isaiah also draws on shepherd/nomadic imagery: "He stretches out the heavens like a curtain, and spreads them like a tent to dwell in" (v. 22), or "He will feed his flock like a shepherd, he will gather the lambs in his arms" (v. 11). Shepherds would speak naturally in terms of tents, curtains, sheep, garden oases, and the simple life of nomads. Farmers and the city-dwellers would speak naturally in terms of foundations, pillars, boundaries, the sedentary life, and cosmic and social order.

What we are given in the first chapters of Genesis are two distinct accounts of creation, the first using the language, imagery, and concerns of the agricultural/urban tradition in Israel, and the second using those of the pastoral/nomadic tradition. This observation helps to explain the inevitable problems in a literal/historical approach to harmonizing the two accounts of creation, as well as, in turn, trying to harmonize both of them with modern scientific accounts of origins. The order of events in the Adam and Eve version in Genesis 2, for example, is quite different from the six-day account with which Genesis begins.

Genesis 1–2:4a	**Genesis 2:4b–24**
(Water and Formless Earth)	(Heavens and Earth Presupposed)
Light (day 1)	Water (mist)
Firmament (day 2)	Adam
Earth and Vegetation (day 3)	Vegetation
Sun, Moon, and Stars (day 4)	Rivers
Fish and Birds (day 5)	Land Animals, Birds (no fish)
Land Animals, Humans (day 6)	Eve

. . . The attempt to interpret these materials as literal, chronological accounts of origins runs into enormous difficulties internally, well before modern scientific scenarios are introduced. Despite valiant efforts by clever exegetes, the two biblical accounts cannot be reconciled, as long as the assumption is made that they are intended to be read as comparable to a natural history. The clue to the differences is to be found within Israel itself where, broadly speaking, there were two main traditions: the pastoral/nomadic and the agricultural/urban. Genesis 1 has drawn upon the imagery and concerns of the farmers and city-dwellers who inhabited river basins prone to flooding, while Genesis 2 has drawn upon the experiences of shepherds, goat-herders, and camel-drivers who lived on the semiarid fringes of the fertile plains, around and between wells and oases. For the pastoral nomads and desert peoples the fundamental threat to life was dryness and barrenness, whereas for those agricultural and urban peoples in or near flood plains the threat was too much water, and the chaotic possibilities of water. It is also revealing that Genesis 2 does not mention a creation of fish, whereas fish in abundance are prominent in Genesis 1 (fish occupy half of day five, with "swarms of living creatures").

This interpretive approach also helps explain why the two versions of creation present such different —nearly opposite—views of human nature. In Genesis 1 human beings are pictured in the lofty terms of royalty, taking dominion over the earth and subduing it—imagery and values drawn from the very pinnacle of ancient civilizations, which Israel itself achieved in the time of Solomon. In Genesis 2, however, Adam and Eve are pictured as *servants* of the garden, living in a garden oasis: essentially the gardener and his wife. And while Genesis 1 refers to humans as made in "the image and likeness of God," in the continuation of the garden story in Genesis 3 the theme of godlikeness is introduced by the *serpent* who tempts Eve with the promise that by eating of the fruit of the Tree of Knowledge, they would be "like God," knowing good and evil. Celebrants of science and technology beware!

Thus, while Genesis 1 is comfortable with the values of civilization and the fruits of its many achievements and creations, Genesis 2 offers an humble view of humanity, a reminder of the simple life and values of the shepherd ancestors, before farming, and even before shepherding, in an Edenic state of food gathering and tending. In this manner these two views of human nature are counterbalanced. They are not contradictory but complementary. Any celebration of human creation and its achievements is tempered by warnings concerning overweening pride and claims to godlikeness. Our heads at times may be in the clouds, but our feet walk on the Earth and are made of clay.

The two accounts of creation in Genesis are contradictory only if taken as literal history, rather than recognizing that they are operating *analogically*, using the contrasting imagery and concerns of the two main traditions in ancient Israel. The biblical accounts are also not in contradiction with modern scientific accounts either, because (again) the biblical accounts are interpreting origins analogically (albeit using very different sets of analogy), not geologically or biologically. To cite Calvin again: "The Holy Spirit had no intention to teach astronomy, and, in proposing instruction meant to be common to the simplest and most uneducated persons, he made use by Moses and the other prophets of popular language, that none might shelter himself under the pretext of obscurity." Thus it may be said that biblical affirmations of creation are in harmony with the science of any age and culture, not because they have been harmonized by clever argument, but because they have little to do with such concerns (pp. 21–24).

Selection excerpted from:

Hyers, Conrad. 2003. Comparing Biblical and Scientific Maps of Origins. In *Perspectives on an Evolving Creation*, edited by K. B. Miller, pp. 19–33. Grand Rapids, MI: Wm. B. Eerdmans. Used by permission.

• •

EVOLUTION AND PURPOSE (MEANINGFULNESS)

If evolution occurred, and the universe was not specially created in its present form by God, does this imply that there is no ultimate purpose or meaning to the universe? Many conservative Christians believe this to be the case. Other Christians, interpreting Scripture differently, believe that the history of the universe is independent of its purpose, which they believe is determined by God, whether or not we know what it is. Humanists hold the position that there is no cosmic or ultimate meaningfulness to the universe, but that individuals may choose to give meaning to their own individual lives.

Both biblical literalists and Christians who are not biblical literalists may be concerned that even if evolution were part of God's plan, and He worked through natural processes, the insertion of such laws between God and creation somehow implies that God is a more distant deity and less involved in human affairs. There is, not surprisingly, a great diversity of opinion within Christian theology on this topic.

Evolution Implies Cosmic Meaninglessness

Proponents of Intelligent Design are concerned with the implications of evolution for cosmic purpose.

The Church of Darwin

. . . The reason the theory of evolution is so controversial is that it is the main scientific prop for scientific naturalism. Students first learn that "evolution is a fact," and then they gradually learn more and more about what that "fact" means. It means that all living things are the product of mindless material forces such as chemical laws, natural selection, and random variation. So God is totally out of the picture, and humans (like everything else) are the accidental product of a purposeless universe. Do you wonder why a lot of people suspect that these claims go far beyond the available evidence?

Selection excerpted from:

Johnson, Phillip E. *Wall Street Journal*, August 16, p. A14.

Nobel Laureate physicist Steven Weinberg holds that evolution indeed implies that the cosmos is purposeless and lacks ultimate meaning. As a humanist, however, he finds meaning in human curiosity and inventiveness.

Epilogue: The Prospect Ahead

. . . [W]hichever cosmological model proves correct, there is not much of comfort in any of this. It is almost irresistible for humans to believe that we have some special relation to the universe, that human life is not just a more-or-less farcical outcome of a chain of accidents reaching back to the first three minutes, but that we were somehow built in from the beginning. As I write this I happen to be in an airplane at 30,000 feet, flying over Wyoming en route home from San Francisco to Boston. Below, the earth looks very soft and comfortable—fluffy clouds here and there, snow turning pink as the sun sets, roads stretching straight across the country from one town to another. It is very hard to realize that all this is just a tiny part of an overwhelmingly hostile universe. It is even harder to realize that this present universe has evolved from an unspeakably unfamiliar early condition, and faces a future extinction of endless cold or intolerable heat. The more the universe seems comprehensible, the more it also seems pointless.

But if there is no solace in the fruits of our research, there is at least some consolation in the research itself. Men and women are not content to comfort themselves with tales of gods and giants, or to confine their thoughts to the daily affairs of life; they also build telescopes and satellites and accelerators, and sit at their desks for endless hours working out the meaning of the data they gather. The effort to understand the universe is one of the very few things that lifts human life a little above the level of farce, and gives it some of the grace of tragedy (pp. 143–144).

Selection excerpted from:

Weinberg, Steven. 1997. *The First Three Minutes*, Basic Book ed. New York: Bantam Books.

Evolution Does Not Imply Cosmic Meaninglessness

John F. Haught is a Catholic theologian at Georgetown University. He presents a view held by many Catholic and Protestant theologians, that the essence of Christianity (the redemption of mankind by the sacrifice of Jesus on the cross) implies a specific view of God that is compatible with an evolving, rather than a specially created, static, creation.

Evolution, Tragedy, and Cosmic Purpose

At the end of his important book, *The First Three Minutes*, physicist Steven Weinberg remarks grimly that the more comprehensible the universe has become to modern science, the more "pointless" it all seems. Many other scientists would agree. Alan Lightman has collected several of their reactions to Weinberg's oft-repeated claim. Astronomer Sandra Faber, for example, states that the universe is "completely pointless from a human perspective" (p. 105).

. . . I have been arguing that, from the point of view of Christian theology at least, canvassing nature for evidence of a divine "plan" distracts us from engaging in sufficiently substantive conversation with evolutionary science. Evolutionists have told a story about nature that is extremely difficult to square with the notion of a divine "plan." Nature's carefree discarding of the weak, its tolerating so much struggle and waste during several billion years of life's history on Earth, has made simplistic portraits of a divinely designed universe seem quite unbelievable.

To admit this much, however, is not necessarily to conclude that there is no "point" or purpose to the universe. We must remember that science as such is not equipped, methodologically speaking, to tell us whether there is or is not any "point" to the universe. If scientists undertake nevertheless to hold forth on such matters, they must admit in all candor that their ruminations are not scientific declarations but at best declarations *about* science.

. . . Thus, any respectable argument that evolution makes the universe pointless would have to be erected on grounds other than those that science itself can provide. And yet, even though science cannot decide by itself the question of whether religious hope is less realistic than cosmic pessimism, we must admit that any beliefs we may hold about the universe, whether pessimistic or otherwise, cannot expect to draw serious attention today unless we can at least display their consonance with evolutionary science. We must be able to show that the visions of hope at the heart of the Abrahamic religious traditions provide a coherent metaphysical backdrop for the important discoveries of modern science (pp. 106–107).

. . . For Christian theology, this would mean seeking to understand the natural world, and especially its evolutionary character, in terms of the outpouring of compassion and the corresponding sense of world renewal associated with the God of Jesus the crucified and risen Christ.

Christians have discerned in the "Christ-event" [the crucifixion] the decisive self-emptying or *kenosis* of God. And at the same time they have experienced in this event a God whose effectiveness takes the form of a power of renewal that opens the world to a fresh and unexpected future. As a Christian theologian, therefore, when I reflect on the relationship of evolutionary science to religion, I am obliged to think of God as both *kenotic love* and *power of the future*. This sense of God as a self-humbling love that opens up a new future for the world took shape in Christian consciousness only in association with the "Christ-event"; and so, as we ponder the implications of such discoveries as those associated with evolution, it would be disingenuous of Christian theologians to suppress the specific features of their own faith community's experience of divine mystery. This means quite simply that in its quest to understand the scientific story of life, Christian theology must ask how evolution might make sense when situated in a universe shaped by God's kenotic compassion and an accompanying promise of new creation (p. 110).

Selection excerpted from:

Haught, John F. 2000. *God After Darwin: A Theology of Evolution*. Boulder, CO: Westview Press.

Hewlett and Peters, a scientist and a theologian, respectively, consider the idea of progress as part of their reflection upon purpose, but conceive of purpose as a much broader concern.

The Problem of Purpose

The central problem faced by advocates of a theistic interpretation of evolution is not one of reconciling science with a literal reading of Genesis. Nor is the problem one of transformation of one species into another. Nor is it a problem raised by gradual change over long periods of deep time. Rather, it is the problem of purpose.

. . . Why is purpose a problem? Can we not think of nature progressing through evolution just as the design of automobiles is progressing through engineering? After all, we speak of technology "evolving," so is it not appropriate to equate evolution with advance? Evolution makes things better, right?

. . . This cultural interpretation of natural selection leads us to say: yes, evolution is a form of progress. As a doctrine of progress, evolution becomes one more modern Western ideology among others. As a doctrine of progress it is a philosophy, a value system, an ideology, maybe even a materialistic religion.

However, if we seek what is narrowly scientific and treat evolution strictly as a theory to explain biological change, then we must expunge all references to purpose. No such thing as progress can be admitted. This is a principle of scientific research which is dogmatic to today's evolutionary biologists. They appeal to Charles Darwin himself for having set the precedent by relying upon chance variation. Natural selection, suggested Darwin, is not a secular form of divine providence. . . . Natural selection favors the fit, to be sure; but what determines fitness has nothing to do with an overall purpose or direction in nature.

. . . This expunging of purpose from nature gives nightmares to theologians. How can we speak of a creation without purpose? How can we speak of redemption without a goal toward which the creation aspires?

. . . On the one hand, people of faith simply cannot conceive of the natural world without purpose or at least value. To be sure, theologians have no investment in the secular doctrine of progress; nor do they feel obligated to use the language of "evolution" to indicate advancement toward a better and better world. Dropping the idea of progress from the long story of nature is no loss, theologically speaking. Yet, on the other hand, giving up totally on purpose merely to satisfy the scientific method seems like a high price to pay. A purposeless creation would not be a creation at all.

So, alas, what's a theologian to do? One option would be to throw in the towel and become a deist. No purpose within nature is discernible because, according to this option, it simply is not there. The deistic theologian could affirm that God created the initial conditions of law and chance that made evolution possible; and then God left the natural world to run itself ever since. If such deism is unsatisfying, another option would be to mix in a divine plot to the scientist's story of nature. The theologian could rewrite the Epic of Evolution by expanding on the story told by scientists. The theologian could declare that this evolutionary story has had a plot all along. When God created the world in the beginning, according to this option, God placed a potential into the creation which now through evolution is becoming actualized. The problem with this Epic of Evolution approach is that it is a dogmatic superimposition; it does not derive from the science itself. A third option would be to treat creation eschatologically—that is, to locate the world's purpose in God rather than in nature. The problem for theologians as posed by scientists is that no purpose can be seen *within* nature. According to this option, nature's purpose is not inherent within nature itself; rather, its value or direction belongs to the relationship of nature with God. God's redemptive vision becomes the source of the divine declaration that nature is "very good."

Selection excerpted from:

Peters, Ted, and Martinez Hewlett. 2003. *Evolution from Creation to New Creation*. Nashville, TN: Abingdon Press. Used by permission.

Chapter 12

The Nature of Science

INTRODUCTION

The creationism/evolution controversy concerns issues of religion, politics, culture—and, of course, science, an institution with considerable influence and authority in American society. Both the antievolutionists and the defenders of evolution seek the imprimatur of science to promote their respective positions. Both sides seek to define science and determine its application, with factions differing strikingly on what they consider to be "good" or acceptable scientific practice.

This chapter will consider three "nature of science" themes that recur in the literature of the creationism/evolution controversy. The first concerns the use of two terms discussed in chapter 1, "fact" and "theory." Evolution, we hear from antievolutionists, is "only a theory" and should not be presented as "fact." Of course evolution is a theory, retort the evolutionary scientists. Theories are much more important than facts! Antievolutionists respond that even if evolution is a theory in the scientific sense (of explanation), it isn't a very *good* theory, and isn't supported by the evidence.

The second topic in this chapter is whether there should be different rules for how science works, depending on the type of question being asked. Antievolutionists suggest it is by distinguishing "origins science" from "operation science"; evolutionary scientists, while drawing a somewhat similar distinction between "historical sciences" and "experimental sciences," maintain that both are scientifically legitimate and are not fundamentally different. This distinction is related to the third topic of this chapter, the relationship between methodological and philosophical naturalism. Antievolutionists, especially Intelligent Design (ID) supporters, believe that evolution seems plausible only to those in the grips of a nonscientific ideology: philosophical naturalism. Evolutionists respond that the sort of naturalism implicit in science is only methodological, adopted not dogmatically but to enable scientific progress.

FACT VS. THEORY

Creationist Views of the Fact/Theory Question

As discussed in chapter 1, fact and theory are used differently in science than on the street. For example, theories are not "proven" but are corroborated, rejected, or modified in light of new data and how well they agree with accepted theories and principles. In many examples of legislation or policies passed by school boards, these terms are used in their vernacular senses rather than in their scientific senses.

"Theory Not Fact" Resolutions and Legislation

A classic example of evolution as theory (meaning "guess" or "hunch") is found in the disclaimer that the Alabama Board of Education ordered placed in biology textbooks from about 1996 until 2002. The disclaimer in its full form is presented in chapter 9. The first two sentences, however, exemplify the idea that "theories" are somehow suspect and not as reliable as "facts." Other legislation or resolutions from other states similarly illustrate nonscientific usage of these terms.

The Alabama Disclaimer, 1995

This textbook discusses evolution, a controversial theory some scientists present as a scientific explanation for the origin of living things, such as plants, animals, and humans. No one was present when life first appeared on earth. Therefore, any statement about life's origins should be considered as theory, not fact. . . .

Columbia County, Georgia, Resolution, 1996 (failed)

. . . The teaching of science should distinguish between theory and fact. Scientific hypotheses which cannot be proven or replicated, such as the theory of evolution, must always be taught as theories and not fact. . . .

Mississippi HB No. 1397, 2003 (failed)

BE IT ENACTED BY THE LEGISLATURE OF THE STATE OF MISSISSIPPI:

SECTION 1. Beginning with the 2004–2005 school year, the State Board of Education shall require that any textbook that includes the teaching of evolution in its contents shall have the following language inserted on the inside front cover of those textbooks: "The word 'theory' has many meanings: systematically organized knowledge, abstract reasoning, a speculative idea or plan, or a systematic statement of principles. Scientific theories are based on both observations of the natural world and assumptions about the natural world. They are always subject to change in view of new and confirmed observations.

"This textbook discusses evolution, a controversial theory some scientists present as a scientific explanation for the origin of living things. No one was present when life first appeared on earth. Therefore, any statement about life's origins should be considered a theory. . . ." [The legislation cites the remainder of the Alabama disclaimer.]

Tennessee SB 3229, 1996 (failed)

. . . BE IT ENACTED BY THE GENERAL ASSEMBLY OF THE STATE OF TENNESSEE: SECTION 1. Tennessee Code Annotated, Title 49, Chapter 6, part 10, is amended by adding the following new section thereto:

. . . No teacher or administrator in a local education agency shall teach the theory of evolution except as a scientific theory. Any teacher or administrator teaching such theory as fact commits insubordination, as defined in Section 49-5-501(s)(6), and shall be dismissed or suspended as provided in Section 49-5-511. . . .

For those antievolutionists who do understand that theories are more than just guesses, evolution nonetheless does not qualify as an acceptable, valid theory.

Darwin's Leap of Faith

. . . In essence, the theory of evolution should never have been accepted as a legitimate scientific theory. True science uses logical inference from empirical observations to arrive at truth (Bird 1991, vol. 2: 14). What the proper definition of science and description of the scientific method will indicate is that, while scientists who study nature may utilize the scientific method, evolutionary theory *itself* is not ultimately good science because evolution has few, if any, "demonstrated truths" or "observed facts." It does not arrive at truth inductively, i.e., the supporting of a general truth (evolution) by observing particular cases that exist (Bird 1991, vol. 1: 15–16). A strictly limited change within species can be demonstrated, such as crossbreeding among some plants and animals, but this has nothing to do with evolution as commonly understood, nor, when examined critically, can the mechanisms involved explain how evolution might occur.

Let's further explain why we do not think the term theory is appropriately applied to evolution. As noted, in science, the term has a more profound meaning than in common usage. In science, the phrase, "It's *just* a theory" is inappropriate. A good scientific theory explains a great deal of scientific knowledge, including "both laws and the facts dependent on scientific laws" (Broad and Wade 1982).

> "Theory"—to a scientist—is a concept firmly grounded in and based upon facts, contrary to the popular opinion that it is a hazy notion of undocumented hypothesis. Theories do not become facts; they explain facts. A theory must be verifiable. If evidence is found that contradicts the stated theory, the theory must be modified or discarded. (Matsumura 1995: 119)

Evolution is not verifiable; it explains few facts, and it contradicts several scientific laws. Further, there is a great deal of undeniable evidence against it. So, scientifically, it must be modified or discarded. It is inappropriate to apply the term *scientific theory* to evolution. Applied to evolution, the term actually fits the popular idea much better, e.g., "a hazy notion or undocumented hypothesis."

REFERENCES CITED

Bird, W. R. 1991. *The Origin of Species Revisited*. 2 vols., *The Theories of Evolution and of Abrupt Appearance*. Cambridge, Ont.: Thomas Nelson.

Broad, W., and Nicholas Wade. 1982. *Betrayers of the Truth: Fraud and Deceit in the Halls of Science*. New York: Simon & Schuster.

Matsumura, M. 1995. *Voices for Evolution*. Berkeley, CA: National Center for Science Education. See New Orleans Geological Society statement, p. 119.

Selection excerpted from:

Ankerberg, John, and John Weldon. 1998. *Darwin's Leap of Faith*, pp. 52–54. Eugene, OR: Harvest House.

Responses to Creationist Views of the Fact/Theory Question

Representative Rush D. Holt is one of the few members of Congress to hold a Ph.D. in science. From this perspective, he criticizes the use of the term "theory" in the "Santorum amendment," a part of the conference committee report of the 2002 education bill passed by Congress (discussed in chapter 9).

Speech of Hon. Rush D. Holt

[Mr. HOLT.] Mr. Speaker, I rise today to address my colleagues regarding H.R. 1, No Child Left Behind. Although we passed this important legislation last week, I must express my reservations about certain language included in the conference report:

> The conferees recognize that a quality science education should prepare students to distinguish the data and testable theories of science from the religious or philosophical claims that are made in the name of science. Where topics are taught that may generate controversy (such as biological evolution), the curriculum should help students to understand the full range of scientific views that exist, why such topics may generate controversy, and how scientific discoveries can profoundly affect society.

Outside of the scientific community, the word "theory" is used to refer to a speculation or guess that is based on limited information or knowledge. Among scientists, however, a theory is not a speculation or guess, but a logical explanation of a collection of experimental data. Thus, the theory of evolution is not controversial among scientists. It is an experimentally tested theory that is accepted by an overwhelming majority of scientists, both in the life sciences and the physical sciences. The implication in this language that there are other scientific alternatives to evolution represents a veiled attempt to introduce creationism—and, thus, religion—into our schools. Why else would the language be included at all? In fact, this objectionable language was written by proponents of an idea known as "intelligent design." This concept, which could also be called "stealth creationism," suggests that the only plausible explanation for complex life forms is design by an intelligent agent. This concept is religion masquerading as science. Scientific concepts can be tested; intelligent design can never be tested. This is not science, and it should not be taught in our public schools.

Mr. Speaker, I am a religious person. I take my religion seriously and feel it deeply. My point here is not to attack or diminish religion in any way. My point is to make clear that religion is not science and science is not religion. The language in this bill can result in diminishing both science and religion.

Selection excerpted from:

Holt, Rush. 2001. Conference Report on HR 1, No Child Left Behind Act of 2001, December 13, 2001. *The Congressional Record*, December 20 (Extensions), p. E2365. See also http://frwebgate.access.gpo.gov/cgi-bin/getpage.cgi. accessed Sept. 30, 2003.

What Is a "Fact" and What Is a "Theory"?

A fact is a confirmed observation. For example, it is a confirmed observation that every tetrapod known has at some stage of its life, a humerus, a radius and ulna, and a distal cluster of bones corresponding to carpals, metacarpals and phalanges. The general public (and even some scientists) use the word "fact" to imply capital T "Truth": unchanging agreement. In science, facts, like theories, may change: it was once a fact (for about 10 years) that *Homo sapiens* had 48 chromosomes. But other observations were confirmed and explanations found for the erroneous observations, and now we know that there are 46. In general, though, in science we treat facts as statements we don't need to test and question anymore, but rather can use as givens to build more complex understandings.

A theory, in science, is a logical construct of facts and hypotheses that attempts to explain a natural phenomenon. It is an explanation, not a guess or hunch that one can casually disregard. Theory formation—explanation—is the goal of science, and nothing we do is more important. A scientist joked that we should applaud the Tennessee law punishing teachers for teaching evolution as a "fact rather than a theory" because "everyone knows that theories are more important than facts!" Theories explain facts, but the general public doesn't know that.

Concerning evolution, then, what's a fact and what's a theory? One hears from many scientists, "Evolution is FACT!!!" The meaning here is that evolution, the "what happened," is so well supported that we don't argue about it, anymore than we argue about heliocentrism versus geocentrism. We accept that change through time happened, and go on to try to explain how. What we mean and what is heard is often different, however. What the public often hears when scientists say "Evolution is FACT!" is that we treat evolution as unchallengeable dogma, which it isn't.

We must learn to present evolution not as "a fact" in this dogmatic sense, but "matter of factly," as we would present heliocentrism and gravitation. Most people consider heliocentrism and gravitation as "facts," but they are not "facts" in my definition of "confirmed observations." Instead, they are powerful inferences from many observations, which are not in themselves questioned, but used to build more detailed understandings.

From the standpoint of philosophy of science, the "facts of evolution" are things like the anatomical structural homologies such as the tetrapod forelimb, or the biochemical homologies of cross species protein and DNA comparisons, or the biogeographical distribution of plants and animals. The "facts of evolution" are observations, confirmed over and over, such as the presence and/or absence of particular fossils in particular strata of the geological column (one never finds mammals in the Devonian, for example). From these confirmed observations we develop an explanation, an inference, that what explains all of these facts is that species have had histories, and that descent with modification has taken place. Evolution is thus a theory, and one of the most powerful theories in science.

We may also speak of "theories" (plural) of evolution, in the sense of the explanations for how descent with modification has taken place. It is conceptually sound to separate evolution as something that did or did not happen from explanations about how, or how fast, or which species are related to which. . . .

Indeed, teachers have to be sure that students know what theories are and why they are important. Students also must—this is crucial—learn as part of their science instruction that our explanations change with new data or better ways of looking at things. Anti-evolutionists make the statement that "evolution isn't science because you guys are always changing your minds about stuff." This is not a criticism. That's the way a vigorous science works.

Selection excerpted from:

Scott, Eugenie C. 1996. Dealing with Antievolutionism. In *Learning from the Fossil Record*, edited by J. Scotchmoor and F. K. McKinney. Pittsburgh, PA: The Paleontological Society. Used with permission.

• •

ORIGINS SCIENCE AND OPERATION SCIENCE

Antievolutionists contend that there are two different kinds of science, dealing with different kinds of problems. Most science deals with the ongoing, regular operations of the natural world, but some scientific questions involve singular events that are not repeated. Many antievolutionists believe that when one is attempting to explain singular events—such as the origin of life or the evolution of some species—it is legitimate to introduce a supernatural agent. Whether to include the actions of a supernatural agent as part of the way of knowing called science is also relevant to the third topic in this chapter, the distinction between philosophical and methodological naturalism.

Two Kinds of Science? Creationist Views

Our intent was to present a selection ("Operation Science and the God Hypothesis") from one of the founding documents of the Intelligent Design movement, *The Mystery of Life's Origin* (Charles B. Thaxton, Walter L. Bradley, Roger L. Olsen. New York: Philosophical Library, 1984), but the authors denied us permission for the excerpt. Readers are encouraged to consult pages 202–205 of that book, or a book by N. L. Geisler and J. K. Anderson expressing the same ideas from a Young Earth Creationist perspective (*Origin Science: A Proposal for the Creation-Evolution Controversy*. Grand Rapids, MI: Baker Book House, 1987).

Thaxton et al. begin by defining science as having the ability (1) to explain what has been observed; (2) to explain what has not yet been observed; and (3) to be tested by further experimentation and modification. They claim that this approach works only if there are recurring events to test theories against, and since these natural events recur during the operation of the universe, this kind of science is known as "operation science." Appealing to God in operation science is illegitimate, because "by definition God's supernatural action would be willed at His pleasure and not in a recurring manner" (p. 202). They argue against "God of the gap" explanations (see chapter 6), where God's hand is invoked to fill a gap in our scientific knowledge, and note how Newton made this error in explaining the rotation of planets around the sun.

Origin science, on the other hand, refers to attempts to understand singular events such as the origin of life. The authors claim that theories about such events cannot be falsified because they occur only once. They contend that because of the differences between origin science and operation science, it is legitimate to allow divine explanations in the former if not the latter.

> There are significant and far-ranging consequences in the failure to perceive the legitimate distinction between origin science and operation science. Without the distinction we inevitably lump origin and operation questions together as if answers to both are sought in the same manner and can be equally known. Then, following the accepted practice of omitting appeals to divine action in recurrent nature, we extend it to origin questions too. The blurring of these two categories partially explains the widely held view that a divine origin of life must not be admitted into the *scientific* discussion, lest it undermine the motive to inquire and thus imperil the scientific enterprise. (pp. 204–205)

> . . . The perception of a threat to scientific inquiry and the possible end of science are legitimate concerns. But we question whether the God-hypothesis in origin science would necessarily have this disastrous effect. . . . In our view, as long as one acknowledges and abides by the above distinction between origin science and operation science, there is no *necessary* reason that Special Creation would have the disastrous effects predicted for it. One must be careful, however, to follow the tradition of early modern scientists and *disallow* any divine intervention in operation science. (p. 205)

Two Kinds of Science? Opposing Views

Ernst Mayr is considered the dean of evolutionary biologists. One of the architects of the modern synthetic theory of evolution in the mid-twentieth century, at the time of the publication of this book he is in his late nineties, and still writing. His view is that although explanation in the historical sciences is importantly different from explanation in the experimental sciences, it is not suspect or illegitimate on that account.

How Does Biology Explain the Living World?

When a biologist tries to answer a question about a unique occurrence such as "Why are there no hummingbirds in the Old World?" or "Where did the species *Homo sapiens* originate?" he cannot rely on universal laws. The biologist has to study all the known facts relating to the particular problem, infer all sorts of consequences from the reconstructed constellations of factors, and then attempt to construct a scenario that would explain the observed facts of this particular case. In other words, he constructs a historical narrative.

Because this approach is so fundamentally different from the causal-law explanations, the classical philosophers of science—coming from logic, mathematics, or the physical sciences—considered it quite inadmissible. However, recent authors have vigorously refuted the narrowness of the classical view and have shown not only that the historical-narrative approach is valid but also that it is perhaps the only scientifically and philosophically valid approach in the explanation of unique occurrences.

It is, of course, never possible to prove categorically that a historical narrative is "true." The more complex a system is with which a given science works, the more interactions there are within the system, and these interactions very often cannot be determined by observation but can only be inferred. The nature of such inference is likely to depend on the background and the previous experience of the interpreter; and therefore, not surprisingly, controversies over the "best" explanation frequently occur. Yet every narrative is open to falsification and can be tested again and again.

For instance, the demise of the dinosaurs was once attributed to the occurrence of a devastating disease to which they were particularly vulnerable, or to a drastic change of climate caused by geological events. Neither assumption was supported by credible evidence, however, and both ran into other difficulties. Yet, when in 1980 the asteroid theory was proposed by Walter Alvarez and, particularly, after the presumed impact crater was discovered in Yucatan, all previous theories were abandoned, since the new facts fit the scenario so well.

Among the sciences in which historical narratives play an important role are cosmogony (the study of the origin of the universe), geology, paleontology, phylogeny, biogeography, and other parts of evolutionary biology. All these fields are characterized by unique phenomena. Every living species is unique and so is, genetically speaking, every individual. But uniqueness is not limited to the world of life. Each of the nine planets of the solar system is unique. On earth, every river system and every mountain range has unique characteristics.

Unique phenomena have long frustrated the philosopher. Hume noted that "science cannot say anything satisfactory about the cause of any genuinely singular phenomenon." He was correct if he had in mind that unique events cannot be fully explained by causal laws. However, if we enlarge the methodology of science to include historical narratives, we can often explain unique events rather satisfactorily, and sometimes even make testable predictions.

The reason why historical narratives have explanatory value is that earlier events in a historical sequence usually make a causal contribution to later events. For instance, the extinction of the dinosaurs at the end of the Cretaceous vacated a large number of ecological niches and thus set the stage for the spectacular radiation of the mammals during the Paleocene and Eocene, owing to their invasion of these vacant niches. The most important objective of a historical narrative is to discover causal factors that contributed to the occurrence of later events in a historical sequence. The establishment of historical narratives does not in the least mean the abandonment of causality, arrived at strictly empirically (pp. 64–66).

Selection excerpted from:

Mayr, Ernst. 1998. *This Is Biology: The Science of the Living World.* Cambridge, MA: Belknap Press.

Creationism, Ideology, and Science

One largely old-earth creationist proposal is that there are two different kinds of science: "operation science," and "origins science" (Thaxton et al., 1984; Geisler and Anderson, 1987). A distinction is made between phenomena which occur "with regularity" and those which occur "singularly." Regularly-occurring phenomena can be studied in the fashion most of us associate with normal science, or "operation science." But one-time phenomena, such the Big Bang, and other evolutionary events comprise what creationists call "origins science."

Of course there are differences in the study of repeatable events vs. non-repeatable ones, but mainstream philosophers of science agree that phenomena of historical sciences like geology, paleontology, and astronomy can be studied scientifically, and even experimentally. Mount St. Helens erupted as a singular event, but this does not prevent there being a science of volcanoes. Similarly, even if bears and dogs split from a common ancestor only once, we can still evaluate the hypothesis that bears and dogs are closely related against empirical evidence (from fossils, comparative anatomy, biochemistry, etc.). We can also learn about the processes that influence evolution by looking at the evidence for other such splits. There are many ways to scientifically study events of this type.

Creationists add an additional factor to this bimodal division of the scientific world, which I believe sheds light on why the division was invented in the first place: it allows the intrusion of the supernatural into scientific explanation. Geisler proposes that to accompany the two kinds of science, there are two kinds of causation: primary causes and secondary causes. Operation science relies properly on secondary causes, but origins science is allowed to invoke primary causes. Thaxton et al. refer to primary cause more bluntly as the "God hypothesis," and agree that in operation science, "the appeal to God is quite illegitimate, since by definition God's supernatural action would be willed at His pleasure and not in a recurring manner" (Thaxton et al., 1984: 203). But when dealing with "origins science," it is not only permissible, but essential, to allow recourse to supernatural causation (i.e., miracles).

Few would argue with not resorting to miracles in operation science, but proponents of this artificial division do not make a solid case for resorting to miracles in origin science. Arguably, non-recurrent events may be more difficult and challenging to study than repeated events, but that in itself is insufficient to require resorting to the supernatural . . . (p. 514).

REFERENCES CITED

Geisler, Norman L., and J. Kerby Anderson. 1987. *Origin Science: A Proposal for the Creation–Evolution Controversy*. Grand Rapids, MI: Baker Book House.

Thaxton, Charles B., Walter L. Bradley, and Roger L. Olsen. 1984. *The Mystery of Life's Origin: Reassessing Current Theories*. New York: Philosophical Library.

Selection excerpted from:

Scott, Eugenie C. 1996. Creationism, Ideology, and Science. In *The Flight from Science and Reason*, edited by P. Gross, N. Levitt and M. W. Lewis. New York: New York Academy of Sciences.

PHILOSOPHICAL AND METHODOLOGICAL NATURALISM

In chapter 3 you read about philosophical naturalism—the idea that material causes (matter, energy, and their interaction) are all there is in the universe: there is no God or gods, nor any supernatural forces. This philosophical idea is contrasted with methodological naturalism, a rule of science (or at least a habit of scientists) that restricts science to explaining the natural world through natural processes. Creationists, especially ID proponents, contend that this is too limiting; science would produce "truer" answers if scientists were allowed to invoke supernatural causation. You can see that this controversy is a cousin to the issue of whether origin science and operation science are valid distinctions; both issues involve the question of whether it is legitimate for science (as a way of knowing) to allow the hand of God as an explanation. Another question is whether methodological naturalism implies or entails philosophical naturalism; if one decides to use only natural explanations in one's science, does this imply a philosophical conclusion that there *can be* no divine involvement in the universe? Here are contrasting views.

Philosophical vs. Methodological Naturalism: Creationist Views

Phillip E. Johnson is one of the most prolific spokespersons for the Intelligent Design form of creationism. He has written extensively on the association of evolution and "Darwinism" with naturalism in the philosophical sense.

Position Paper on Darwinism

. . . 1.0 The important issue is not the relationship of science and creationism, but the relationship of science and materialist philosophy.

. . . 1.2 . . . The question I raise is not whether science should be forced to share the stage with some Biblically based rival known as creationism, but whether we ought to be distinguishing between the doctrines of scientific materialist philosophy and the conclusions that can legitimately be drawn from the empirical research methods employed in the natural sciences.

1.3 Scientific materialism (or naturalism . . .) is the philosophical doctrine that everything real has a material basis, that the path to objective knowledge (distinct from subjective belief) is exclusively through the methods of investigation accepted by the natural sciences, and that teleological conceptions of nature ("we are here for a purpose") are invalid. To a scientific materialist there can be no "ghost in the machine," no non-material intelligence which created the first life or guided its development into complex form, and no reality which is in principle inaccessible to scientific investigation, i.e., supernatural.

1.4 The metaphysical assumptions of scientific materialism are not themselves established by scientific investigation, but rather are held a priori as unchallengeable and usually unexamined components of the "scientific" world view. . . . The naturalistic evolution of life from prebiotic chemicals and its subsequent naturalistic evolution into complexity and humanity is assumed as a matter of first principle, and the only question open to investigation is how this naturalistic process occurred.

1.5 The question is whether this refusal to consider any but naturalistic explanations has led to distortions in the interpretation of empirical evidence, and especially to claims of knowledge with respect to matters about which natural science is in fact profoundly ignorant. . . .

2.0 The continued dominance of neo-Darwinism is the most important example of distortion and overconfidence resulting from the influence of scientific materialist philosophy upon the interpretation of the empirical evidence. . . .

3.0 The refusal (or inability) of the scientific establishment to acknowledge that Darwinism is in serious evidential difficulties and probably false as a general theory is due to the influence of scientific materialist philosophy and certain arbitrary modes of thought that have become associated with the scientific method.

3.1 Science requires a paradigm or organizing set of principles and Darwinism has fulfilled this function for more than a century. It is the grand organizing theoretical principle for biology—a statement which does not imply that it is true.

3.2 Once established as orthodox, a paradigm customarily is not discarded until it can be replaced with a new and better paradigm which is acceptable to the scientific community. Disconfirming evidence (anomalies) can always be classified as "unsolved problems," and the situation remains satisfactory for researchers because even an inadequate paradigm can generate an agenda for research.

3.3 To be acceptable a paradigm must conform to the philosophical tenets of scientific materialism. For example, the hypothesis that biological complexity is the product of some preexisting creative intelligence or vital force is not acceptable to scientific materialists. They do not fairly consider this hypothesis and then reject it as contrary to the evidence; rather they disregard it as inherently ineligible for consideration.

3.4 Given the above premises, something very much like Darwinism simply must be accepted as a matter of logical deduction, regardless of the state of the evidence. Random mutation and natural selection must be credited with shaping biological complexity, because nothing else could have been available to do the job. . . . Because the escape from Darwinism seems to lead nowhere, Darwinism for scientific materialists is inescapable. . . .

5.0. The important debate is not between "evolutionists" and "creationists," but between Darwinists (i.e., scientific materialists) and persons who believe that purely naturalistic or materialistic processes may not be adequate to account for the origin and development of life.

5.1 Once separated from its materialistic-mechanistic basis in Darwinism, "evolution" is too vague a concept to be either true or false. If I am told that the phyla of the Cambrian explosion evolved in some non-Darwinian sense from preexisting bacteria or algae, I do not know what the claim adds to the simple factual statement that the prokaryotes came first. It conveys no information about how the new forms came into existence, and the "evolution" in question could be something as metaphysical as the evolution of an idea in the mind of God.

5.2 Similarly, whether "creation" occurred over a greater or lesser period of time, or whether new forms were developed from older ones rather than from scratch, is not fundamental. The truly fundamental question is whether the natural world is the product of a preexisting intelligence, and whether we exist for a purpose which we did not invent ourselves. If Darwinists have not been overstating their case, they have disproved the theistic alternative, or at least made consideration of it superfluous. . . .

6.0 Whatever may be its utility as a paradigm within the restrictive conventions of scientific materialism, Darwinism has continually been presented to the public as the factual basis for a comprehensive world view that excludes theism as a possibility. A few representative quotations will suffice to make the point:

6.1 George Gaylord Simpson: "Although many details remain to be worked out, it is already evident that all the objective phenomena of the history of life can be explained by purely naturalistic or, in a proper sense of the sometimes abused word, materialistic factors. They are readily explicable on the basis of differential reproduction in populations (the main factor in the modern conception of natural selection) and of the mainly random interplay of the known processes of heredity. . . . Man is the result of a purposeless and natural process that did not have him in mind." (*The Meaning of Evolution*)

. . . 6.3 Richard Dawkins: "Darwin made it possible to be an intellectually fulfilled atheist." (*The Blind Watchmaker*)

. . . 7.0 Whether the materialist-mechanist program has succeeded as the Darwinists have so vehemently claimed is a legitimate subject for intellectual exploration. Scientists rightly fight to protect their freedom from dogmas that others would impose upon them. They should also be willing to consider fairly the possibility that they have been seduced by a dogma which they found too attractive to resist. . . .

Selection excerpted from:

Johnson, Phillip E. 2003. *Position Paper on Darwinism*. Apologetics.org. Accessed September 30, 2003. Available from http://www.apologetics.org/articles/positionpaper.html.

Philosophical vs. Methodological Naturalism: Scientists and Philosophers Respond

Here are two selections from philosophers.

Methodological Naturalism and Evidence

. . . But is the methodological rule itself dogmatic? . . . Does science put forward the methodological principle not to appeal to supernatural powers or divine agency simply on authority? . . . Certainly not. There is a simple and sound rationale for the principle based upon the requirements of scientific evidence.

Empirical testing relies fundamentally upon use of the lawful regularities of nature that science has been able to discover and sometimes codify in natural laws. For example, telescopic observations implicitly depend upon the laws governing optical phenomena. If we could not rely upon these laws—if, for example, even when under the same conditions, telescopes occasionally magnified properly and at other occasions produced various distortions dependent, say, upon the whims of some supernatural entity—we could not trust telescopic observations as evidence. The same problem would apply to any type of observational data. Lawful regularity is at the very heart of the naturalistic world view and to say that some power is supernatural is, by definition, to say that it can violate natural laws. So, when Johnson argues that science should allow in supernatural powers and intelligences he is in effect saying that it should allow beings that are above the law (a rather strange position for a lawyer to take). But without the constraint of lawful regularity, inductive evidential inference cannot get off the ground. Controlled, repeatable experimentation, for example, which Johnson explicitly endorses in his video "Darwinism on Trial," would not be possible without the methodological assumption that supernatural entities do not intervene to negate lawful natural regularities.

Of course science is based upon a philosophical system, but not one that is extravagant speculation. Science operates by empirical principles of observational testing; hypotheses must be confirmed or disconfirmed by reference to empirical data. One supports a hypothesis by showing consequences obtained that would follow if what is hypothesized were to be so in fact. Darwin spent most of the *Origin of Species* applying this procedure, demonstrating how a wide variety of biological phenomena could have been produced by (and thus explained by) the simple causal processes of the theory. Supernatural theories, on the other hand, can give no guidance about what follows or does not follow from their supernatural components. For instance, nothing definite can be said about the processes that would connect a given effect with the will of the supernatural agent—God may simply say the word and zap anything into or out of existence. Furthermore, in any situation, any pattern (or lack of pattern) of data is compatible with the general hypothesis of a supernatural agent unconstrained by natural law. Because of this feature, supernatural hypotheses remain immune from disconfirmation. Johnson's form of creationism is particularly guilty on this count. Creation-Science does include supernatural views at its core that are not testable and it was rightly dismissed as not being scientific because of these in the Arkansas court case, but it at least was candid about a few specific nonsupernatural claims that are open to disconfirmation (and indeed that have been disconfirmed), such as that the earth is less than 10,000 years old and that many geological and paleontological features were caused by a universal flood (the Noachian Deluge). Johnson, however, does not provide any creationist claim beyond his generic one that "God creates for some purpose," and as a purely supernatural hypothesis this is not open to empirical test. Science assumes Methodological Naturalism because to do otherwise would be to abandon its empirical evidential touchstone (pp. 88–89).

Selection excerpted from:

Pennock, Robert T. 1999. *Tower of Babel: The Evidence Against the New Creationism*. Cambridge, MA: Bradford Book/MIT Press.

Creationist Strategy 1: The Nature of Science

First, what does Johnson mean by "naturalism"? Johnson begins his book, *Reason in the Balance*, by characterizing naturalism as a metaphysical assumption: "the doctrine that nature is all there is," and he goes on to claim that natural science is "based on naturalism." The only content that Johnson gives to the doctrine that "nature is all there is" is atheism. Naturalists, he says, are those who "assume that God exists only as an idea in the minds of religious believers." And: "If naturalism is true, then humankind created God—not the other way around." If naturalism is the assumption that God exists only as an idea in the mind of religious believers, then it is pretty clear that natural science is *not* based on naturalism. It is absurd to suppose that science, which is totally silent on the question of God, is based on the assumption that God exists only as an idea in the mind of religious believers. So, this metaphysical naturalism—whether it is a bias or not—is not an underpinning of science.

But there is another kind of naturalism, *methodological* naturalism, that Johnson discusses in his Appendix. "A methodological naturalist," he says, "defines science as the search for the best naturalistic theories," where a naturalistic theory abjures supernatural causes. This amounts to saying that methodological naturalism is the view that no naturalistic explanation can appeal to God or to any supernatural phenomena. In this sense of methodological naturalism, I would agree with Johnson, science is committed to naturalism. But is methodological naturalism just a bias in science? Is it a bias to exclude supernatural explanations from science?

Surely not. If the methodological naturalism that is a hallmark of science were a mere bias, the explosion of scientific knowledge from the 16th and 17th centuries on would be totally inexplicable. The proof of the pudding is in the eating, and the sciences are unparalleled as generators of knowledge of the natural world. It makes little sense to say that scientists have misunderstood their own enterprise, that they should count as scientific explanations those that appeal to a supernatural being. It makes little sense to rebuke such a successful practice for having the character that it has. What counts as a scientific explanation is determined by science. So, taking methodological naturalism as the view that no explanation that appeals to a Creator or an Intelligent Designer is a scientific explanation, Johnson's charge that methodological naturalism is a *bias* in science is off the mark.

. . . Scientific explanations—explanations put forward on the basis of scientific consideration—are fully naturalistic, and have no place for appeal to a supernatural agent. It does not follow from this, however, that *all* correct explanations are scientific explanations. We must distinguish between scientific claims—claims made from *within* science—and claims made *about* science. One important claim about science (one that I reject) is that science is the arbiter of all knowable truth, that there is nothing to be known beyond what science delivers. Call this claim "scientism."

(Scientism) Science is the arbiter of all knowable truth.

If scientism were correct, then from the commitment of science to methodological naturalism, it would follow that all correct explanations (not just scientific explanations) are naturalistic. That stance would rule out, a priori, any explanation that appealed to God. This, I think, would be a bias. But this does not follow from the methodological naturalism of science; it follows only with the addition of the metaphysical, extra-scientific thesis of scientism. Scientism is like a closure principle—"and that's all there is." If we reject scientism, as I think that we should, then from the fact that all scientific explanations are naturalistic, it does not follow that all legitimate explanations are naturalistic. So, exclusion of God from the science classroom is not necessarily exclusion of God elsewhere—for example, where we are trying to give a metaphysical account of why there is something rather than nothing at all. This latter question—Why is there anything rather than nothing at all?—is not a scientific question and will not be susceptible to a scientific explanation. But unless we are scientistic, we may think that there is some explanation—albeit not a scientific one. Again, however, questions not susceptible of scientific answers do not belong in a science class.

. . . To sum up: Science is not committed to the nonexistence of God, as it would be if it were based on metaphysical naturalism. Science is committed to naturalistic explanations. Science does not count any explanation that appeals to God or to supernatural phenomena as a scientific explanation (thus, it is committed to methodological naturalism). But methodological naturalism is no bias: it is in the nature of science. And unless one conjoins methodological naturalism with scientism, nothing at all follows about the nonexistence of God. So, methodological naturalism (but not scientism) is part of science, and given the success of science, it is idle to charge that science should be something other than what it is (pp. 57–59).

Selection excerpted from:

Baker, Lynn Rudder. 2000. God and Science in the Public Schools. *Philosophic Exchange* 30: 53–69.

Richard Dickerson is a molecular biologist.

The Game of Science

Science, fundamentally, is a game. It is a game with one overriding and defining rule:

> Rule No. 1: Let us see how far and to what extent we can explain the behavior of the physical and material universe in terms of purely physical and material causes, without invoking the supernatural.

Operational science takes no position about the existence or non-existence of the supernatural; it only requires that this factor is not to be invoked in scientific explanations. Calling down special-purpose miracles as explanations constitutes a form of intellectual "cheating." A chess player is perfectly capable of removing his opponent's king physically from the board and smashing it in the midst of a tournament. But this would not make him a chess champion, because the rules had not been followed. A runner may be tempted to take a short-cut across the infield of an oval track in order to cross the finish line ahead of his faster colleague. But he refrains from doing so, as this would not constitute "winning" under the rules of the sport.

Similarly, a scientist also can say to himself, "I believe that *Homo sapiens* was placed on this planet by a special act of divine creation, separate and apart from the rest of living creatures." While this can be a genuinely held private belief, it can never be advanced as a scientific explanation, because once again it violates the rules of the game. If that situation were true, and if *H. sap*. were indeed the result of a special miracle, then, in view of Rule No. 1, above, the only proper scientific assessment would be: "Science has no explanation." The problem with any such statement is that we know from past experience that it probably should have been qualified: "Science has no explanation—yet." As people who have grown up amid the current scientific revolution know full well, last year's miracle is this year's technology.

The vital importance of excluding miracles and divine intervention from the game of science, as is advocated even today by the creationist movement, is that allowing such factors to be invoked as explanations discourages the search for other and more systematic causes. Two centuries ago, if Benjamin Franklin and his contemporaries had been content to regard vitreous and resinous forms of static electricity only as expressions of divine humor, we would be unlikely to have the science of electromagnetism today. A century later, a passive belief that God made all the molecules "after their own kind" would have stunted the infant science of chemistry. And a contemporary who believes devoutly that there are no connections between branches of living organisms is unlikely ever to discover such connections as do exist. The most insidious evil of supernatural creationism is that it stifles curiosity and therefore blunts the intellect.

There are those who demand, in a bizarre misapplication of courtroom standards, that the claims of modern science either be proven beyond a shadow of a doubt at this present moment, or else be given up entirely. Such people do not understand the structure of science as a game. We do not say, "Science absolutely and categorically denies the existence and intervention of the supernatural." Instead, as good game players, we say, "So far, so good. We haven't needed special miracles yet." The particular glory of science is that such an attitude has been so successful, over the past four centuries, in explaining so much of the world around us. A good maxim is: *If it isn't broke, don't fix it.* The game of rational science has been enormously successful. We change the rules of that game at our peril.

To be sure, many areas exist where we as scientists do not yet know all the answers. But these problem areas change from one generation to another, and that which might have seemed miraculous (to some) a generation ago now is seen to be perfectly explicable by natural causes. In hindsight we would have felt foolish had we written off those areas as the result of miracles fifty years ago; and we would be ill-advised to set ourselves up for ridicule by those who will follow us fifty years from now. It is a reasonable prediction that the attitude of future generations toward twentieth-century "scientific creationism" (an inherent oxymoron according to Rule No. 1, above) will be one of ridicule.

It would augur well, for both science and religion, if creationists and evolutionary biologists would realize jointly that the question of the existence or the nonexistence of a Deity is irrelevant to the study of biological evolution. Both the die-hard atheist and the theistic evolutionist can function as modern biologists with absolute integrity. The people who are entirely beyond the pale intellectually are those who can be characterized as short-Earth creationists and Biblical literalists—those who maintain that it all happened in 6 standard 24-hour days, with the celestial equivalent of a wave of a magic wind. A clear line of demarcation must be drawn between such people and evolutionists of either theistic or nontheistic inclination. Some creationist rhetoricians would like to draw the line between nontheistic and theistic evolutionists and to lump the second group (which probably includes the majority of nonscientists) together with the 6-day Young-Earth modern "Know-Nothings." We absolutely must not let them get away with such a tactic.

Science is not a closed body of dogma; it is a continuing process of enquiry. A dry and querulous legalism that tends to inhibit or close off that process is antithetical to science. The cartoonist Sidney Harris once published a cartoon depicting two scientists in consultation before a blackboard filled with equations—obviously some kind of proof in the making. One scientist points to a particular equation and proclaims confidently, "And at this point a miracle occurs!" Real scientists don't talk that way—not because some of them don't believe in miracles, sometime, somewhere—but *because invoking miracles and special creation violates the rules of the game of science and inhibits its progress*. People who do not understand that concept can never be real scientists, and should not be allowed to misrepresent science to young people from whom the ranks of the next generation of scientists will be drawn.

Selection excerpted from:

Dickerson, Richard E. 1992. The Game of Science. *Journal of Molecular Evolution* 34: 277–279.

References for Further Exploration

I hope that you have found the creationism/evolution controversy interesting enough to want to learn more about it than is possible in this small book. Or you may wish to write a paper on one of the many topics touched upon in the previous chapters. This chapter is for you!

Throughout this book, in every chapter, you will have noticed one or more lists of references cited within the chapter. These should of course be consulted for further reading on topics you may find interesting. But there are more references listed in this chapter to guide you beyond the chapter citations.

The topics covered in this chapter parallel the content of the preceding 12 chapters. Both Web sites and printed material (books and articles) can be found, as well as occasional references to video sources. The effort has been made to provide sources that are more easily available in libraries or on the Internet. There are occasional references to scientific journals; these are most easily found in a college or university library. Public libraries may have access to online versions of these journals, but this access will vary from site to site.

The creationism and evolution controversy ranges over a variety of subjects, including science, religion, education, law, politics, and history. In this book, we have barely scratched the surface of these details, much less explored the nuances of the various positions. Those who would like to explore this controversy further are encouraged to consult the following resources.

But first a disclaimer: neither the author nor the publisher vouches for the accuracy of the information in the following books, articles, videotapes, and Web sites. Because positions taken in these resources may be diametrically opposed to one another, not all of them can be correct and accurate; readers are encouraged to read all sources critically. Similarly, reference to a person's or an organization's published work should not be taken as agreement with or endorsement of any of the positions taken by that person or organization.

Most of the books selected were in print or about to be published as of October 2003. All of the Web pages were current as of October 2003: although these Web pages were chosen in part because they are likely to be stable, the Internet is notoriously changeable, and neither the author nor the publisher can guarantee that any particular Web site will continue to exist.

CHAPTER 1: SCIENCE: TRUTH WITHOUT CERTAINTY

John A. Moore (1999) discusses science as a way of knowing in the context of the history of modern biology; Sober (2000) is probably the most accessible textbook on the philosophy of biology. There are many books that discuss the nature of scientific reasoning; one popular introductory text is Giere (2003). American Association for the Advancement of Science (1989) is intended to promote scientific literacy in general; chapter 1 deals with the nature of science. National Academy of Sciences (1998) discusses the nature of science in the particular context of evolution; chapters 3 and 5—"Evolution and the Nature of Science" and "Frequently Asked Questions About Evolution and the Nature of Science"—are particularly relevant to the topic of this chapter. National Academy of Sciences (1999) and chapters 2–4 of Kitcher (1982) are also helpful.

CHAPTER 2: EVOLUTION

A brief treatment of astronomical evolution is Fraknoi et al. (2001). For origins-of-life research, see Wills and Bada (2000) and Bada and Lazcano (2003). Fortey (1998) and Tudge (2000) are two impressive attempts to survey the history of life. On the Web, Maddison (2003) endeavors to chart the phylogenetic tree; American Museum of Natural History (2003) and University of California Museum of Paleontology are the Web sites of two major paleontology collections, and they both also contain materials on evolutionary biology. For both evolution and its mechanisms, the video series *Evolution* (WGBH and Clear Blue Sky 2001), which aired on PBS in 2001, is excellent, as is its companion volume (Zimmer 2001), which additionally provides a good explanation of natural selection. Ruse (2001) discusses controversies in (and about) evolution throughout its history. For human evolution, Johanson and Edgar (1996) is a worthwhile book, and Institute of Human Origins (2003) is worth visiting on the Web. An excellent example of adaptive radiation is the Hawaiian honeycreeper; see Pratt (2004). A classic telling of the most famous adaptive radiation story, the Galapagos finches, is Weiner (1994).

CHAPTER 3: BELIEFS: RELIGION, CREATIONISM, AND NATURALISM

A recent introductory text on science and religion is McGrath (1999); the on-line magazine *Metanexus* provides a forum for relevant discussions. For a scholarly treatment of the creation/evolution continuum, see Scott (1997); for a briefer but more accessible treatment, see Scott (2000). The two major Young Earth Creationist organizations are on the Web at Institute for Creation Research and Answers in Genesis; both sites have ample resources promoting the YEC position. On the ICR web site, see especially the on-line versions of the "Impact" series. Henry M. Morris (1974) is a central YEC text, although there are many others. The Web site Talk.Origins has many refutations of, especially, YEC positions. A good place to begin is Isaak (2003). The major Old Earth Creationist organization is on the Web at Reasons to Believe; Ross (1994) is a representative book by its president and the leading public spokesperson for OEC. The major Intelligent Design Creationist organization is on the Web at Center for Science and Culture; Dembski and Kushiner (2001) is a collection of essays reflecting religious views associated with Intelligent Design Creationism. Keith Miller (2003) is a collection of essays mostly by Protestant theistic evolutionists; Haught (2001), from a Catholic theistic evolutionist, is in a convenient question-and-answer format. Dawkins (1987) is often regarded as a paradigm of atheistic evolutionism.

CHAPTER 4: BEFORE DARWIN TO THE TWENTIETH CENTURY

A classic history of the pre-Darwinian revolution in geology is Gillespie (1996). The best history of the theory of evolution is Bowler (2003), but Ruse (1999) is perhaps a better read. A brief introduction in cartoon form to Darwin's life and times is Miller and Van Loon (1982); among the many excellent biographies of Darwin, Desmond and Moore (1991) is perhaps the most readable, but his autobiography (Darwin 1958) ought not to be skipped. On the Web, British Broadcasting Corporation and WGBH and Clear Blue Sky (2001) are useful. The *Origin of Species* is essential reading, preferably in a facsimile of the first edition (Darwin 1966); it is also worthwhile to sample a range of Darwin's work either through an anthology (Ridley 1996; Appleman 2000) or on the Web at British Library. For the initial responses to the *Origin* in the United States, see Roberts (1988) and Numbers (1998). A definitive treatment of the idea of design is Ruse (2003).

CHAPTER 5: ELIMINATING EVOLUTION, INVENTING CREATION SCIENCE

For Fundamentalism in general, see Marsden (1991). By far the best book on the history of creationism is Numbers (1992). The definitive book on the Scopes trial is the Pulitzer Prize–winning Larson (1997); two useful Web sites on the trial are Linder, administered by a law professor at the University of Missouri, Kansas City, and American Experience, which complements a documentary about the trial. The Institute for Creation Research, the Creation Research Society, and Answers in Genesis are all on the Web at Institute for Creation Research, Creation Research Society, and Answers in Genesis. For the legal history of the creationism/evolution controversy, which largely drove the development of creationism in the period under discussion, see Larson (2003). Transcripts of the opinions in *Epperson v. Arkansas* and *McLean v. Arkansas*, together with links to related material, are available on the Web at Talk.Origins Archive; look for "legal decisions" in the index.

CHAPTER 6: NEOCREATIONISM

A transcript of the opinion in *Edwards v. Aguillard*, together with links to related material, are available on the Web at Talk.Origins Archive; look for "legal decisions" in the index. Larson (2003) is excellent on the legal issues. Although Bird (1987) is out of print, a number of articles are available on-line at Institute for Creation Research. The Foundation for Thought and Ethics is on the Web at Foundation for Thought and Ethics. Many articles by Behe and Dembski are on the Web at Center for Science and Culture and Access Research Network; for criticism, see K. R. Miller (1999), Pennock (2001), and, on the Web, Behe's Empty Box, Talk.Design, and Talk.Origins Archive. A series of point/counterpoint articles between ID proponents Michael Behe and William Dembski and their critics is found on the Web at Boston Review (1996–2003). For articles by Johnson espousing the cultural renewal aspect of Intelligent Design, browse through his articles on the Web at Center for Science and Culture and Access Research Network; Pennock (1999) perceptively criticizes Johnson. Witham (2002) attempts to give the history; Forrest and Gross (2003) is a scathing attack.

CHAPTER 7: COSMOLOGY, ASTRONOMY, GEOLOGY

For Young Earth Creationist perspectives on these topics, consult Henry M. Morris (1974) and the articles available on-line at Answers in Genesis and Institute for Creation Research; it is instructive to compare these with the views of Old Earth Creationists such as Ross (1994) and Reasons to Believe, which are closer to mainstream science. Stenger (2003) is skeptical about the Old Earth Creationist interpretations of cosmology. There is no book aimed specifically at refuting Young Earth Creationist views about cosmology and astronomy, although Plait (2002) devotes a chapter to doing so; Strahler (1999) does an admirable job of refuting Young Earth Creationism views not only about geology but about other disciplines as well. Fraknoi et al. (2001) is good on astronomy, and on the Web, Talk.Origins Archive is a reliable source of accessible refutations of creationism claims about cosmology, astronomy, and geology.

CHAPTER 8: PATTERNS AND PROCESSES OF BIOLOGICAL EVOLUTION

For creationist views on biology, consult Gish (1985), Davis and Kenyon (1993), and (for human evolution in particular) Lubenow (1992). On the Web, try Institute for Creation Research, Answers in Genesis, and Creation Research Society. Many of the readings for chapters 3 and 6 would provide useful background for understanding these views. For refutations from the point of view of mainstream science, Berra (1990) is a basic and readable introduction; Kitcher (1982) and Futuyma (1995) delve further. On the Web, Talk.Origins Archive is a reliable source of accessible refutations of creationism claims about biology; the Fossil Hominids section is especially good for human evolution. Many of the readings for chapters 1, 2, and 4 provide useful background for understanding these refutations. On the Web, Van Till's (2002) analysis of William Dembski's views is useful. Behe's concept of irreducible complexity is discussed in the context of the Design Argument by K. R. Miller (1994).

CHAPTER 9: LEGAL ISSUES

The definitive legal history of the creationism/evolution controversy is Larson (2003), by the author of the definitive history of the Scopes trial (Larson, 1997). Randy Moore (2001) is a legal history by a professor of biology who is also the editor of *The American Biology Teacher*. As always, Talk.Origins Archive has useful information and links; look in the index under "legal decisions." Although Bird (1987) is out of print, a number of Bird's articles addressing the legal issues are available on-line at Institute for Creation Research. Beckwith (2003) argues that it would not be unconstitutional to teach Intelligent Design Creationism in the public schools; a number of articles by Beckwith, David DeWolf, and their colleagues addressing the legal issues are available on-line at Center for Science and Culture. For an opposing view, see Wexler (1997). The Santorum language is discussed in Branch and Scott (2003).

CHAPTER 10: EDUCATIONAL ISSUES

For the creationist point of view, consult Davis and Kenyon (1993) as well as a number of articles available on-line at Center for Science and Culture and Institute for Creation Research. Wells (2000) and articles by Wells that are available on-line at Center for Science and Culture criticize biology textbooks; extensive responses are available on-line at National Center for Science Education and Talk.Origins Archive. Statements supporting the teaching of evolution and opposing the teaching of creationism are collected in Matsumura (1995); section 9 of Pennock (2001) contains a debate between two philosophers on the topic. For teachers, Alters and Alters (2001) is a very useful discussion of the challenges of teaching evolution at the high school level; on the Web, Evolution and the Nature of Science Institutes encourages teachers to teach evolutionary thinking in the context of a more complete understanding of modern scientific thinking. Binder (2002) gives a sociologist's perspective on attempts to introduce creationism into the public schools. The National Academy of Science's book for teachers, *Teaching About Evolution and the Nature of Science* (1998), also deals with issues raised in this chapter. This resource is also on-line at the National Academy Press.

CHAPTER 11: RELIGIOUS ISSUES

A dialogue between scientists interested in religion and theologians interested in science has expanded greatly since the mid-1990s, and so has the literature. Not all of this literature concerns the creationism/evolution issue, but there is an abundance nonetheless. A good place to begin to explore the "science and religion" movement is Russell (2000); then see the Center for Theology and the Natural Sciences' web site, which publishes the journal Science and Theology. The CTNS site is linked to Counterbalance, which has video clips from a number of conferences featuring scientists and theologians. The American Association for the Advancement of Science sponsors a Web site, Dialogue on Science, Ethics, and Religion, which posts science and religion articles and other references. Ian Barbour's treatments of science and religion (1997, 2002) are classic and are reflected in most subsequent scholarship. Haught (2000, 2001, 2003) deals specifically with evolution and Christian theology. The philosopher Michael Ruse (2001) considers science and religion from a compatiblist perspective.

CHAPTER 12: THE NATURE OF SCIENCE

In addition to the readings for chapter 1, Ruse (1996) is a useful collection of papers on the philosophical issues. Although the popular idea that evolution is "just a theory" is widespread, it is more a cultural perspective than a position advocated; for critical discussions, see Gould (1994) and material available on-line at Talk.Origins Archive and WGBH and at Clear Blue Sky (2001). Davis and Kenyon (1993) contains a discussion of evolution as fact and as theory from the point of view of Intelligent Design Creationism. The operations/origins science distinction is frequently invoked in articles on-line at Answers in Genesis. For discussions of methodological and philosophical naturalism, see sections 2 and 5 of Pennock (2001). Articles by Intelligent Design Creationists on naturalism are available on-line at Access Research Network and Center for Science and Culture; look especially for articles by Johnson, Meyer, and Nelson. Pennock (1999) is a good response.

REFERENCES CITED

Alters, Brian J., and Sandra M. Alters. 2001. *Defending Evolution: A Guide to the Evolution/Creation Controversy*. Sudbury, MA: Jones and Bartlett.

American Association for the Advancement of Science. 1989. *Science for All Americans*. New York: Oxford University Press.

Appleman, Philip, ed. 2000. *Darwin*, 3rd ed. New York: Norton.

Bada, Jeffrey L., and Antonio Lazcano. 2003. Prebiotic Soup—Revisiting the Miller Experiment. *Science* 300 (5620):745–746.

Barbour, Ian G. 1997. *Religion and Science: Historical and Contemporary Issues*, rev. ed. San Francisco: HarperSanFrancisco. (Original ed. titled *Religion in an Age of Science*.)

Barbour, Ian G. 2002. *Nature, Human Nature, and God*. Minneapolis: Fortress Press.

Beckwith, Francis J. 2003. *Law, Darwinism & Public Education*. Lanham, MD: Rowman & Littlefield.

Berra, Tim M. 1990. *Evolution and the Myth of Creationism*. Stanford, CA: Stanford University Press.

Binder, Amy J. 2002. *Contentious Curricula: Afrocentrism and Creationism in American Public Schools*. Princeton, NJ: Princeton University Press.

Bird, Wendell. 1987. *The Origin of Species Revisited: The Theories of Evolution and Abrupt Appearance*. Vol. 1. New York: Philosophical Library.

Bowler, Peter J. 2003. *Evolution: The History of an Idea*, 3rd ed. Berkeley: University of California Press.

Branch, Glenn, and Eugenie C. Scott. 2003. The Anti-Evolution Law That Wasn't. *The American Biology Teacher* 65 (3): 165–166.

Darwin, Charles. 1958. *The Autobiography of Charles Darwin*, edited by Nora Barlow. New York: Norton.

Darwin, Charles. 1966. *On the Origin of Species: A Facsimile of the First Edition*. Cambridge, MA: Harvard University Press.

Davis, Percival W., and Dean H. Kenyon. 1993. *Of Pandas and People*, 2nd ed. Dallas, TX: Haughton.

Dawkins, Richard. 1987. *The Blind Watchmaker*. New York: Norton.

Dembski, William A., and James M Kushiner, eds. 2001. *Signs of Intelligence: Understanding Intelligent Design*. Grand Rapids, MI: Brazos Press.

Desmond, Adrian, and James Moore. 1991. *Darwin: The Life of a Tormented Evolutionist*. New York: Warner Books.

DeWolf, David K. 1999 . *Teaching the Origins Controversy: A Guide for the Perplexed*. Discovery Institute, 1999. Accessed November 20, 2002. Available from http://www.discovery.org/viewDB/index.php3?program=CRSC&command= view&id=48.

Forrest, Barbara, and Paul R Gross. 2003. *Creationism's Trojan Horse*. New York: Oxford University Press.

Fortey, Richard. 1998. *Life: A Natural History of the First Four Billion Years of Life on Earth*. New York: Knopf.

Fraknoi, Andrew, George Greenstein, Bruce Partridge, and John Percy. 2001. An Ancient Universe: How Astronomers Know the Vast Scale of Cosmic Time. *The Universe in the Classroom* 56:1–23 (fall). http://www.astrosociety.org/education/publications/tnl/56/.

Futuyma, Douglas J. 1995. *Science on Trial: The Case for Evolution*. Sunderland, MA: Sinauer.

Giere, Ronald N. 2003. *Understanding Scientific Reasoning*, 4th ed. Belmont, CA: Wadsworth.

Gillispie, C. C. 1996. *Genesis and Geology*. Cambridge, MA: Harvard University Press.

Gish, Duane T. 1985. *Evolution: The Fossils Still Say No!* San Diego: Institute for Creation Research.

Gould, Stephen Jay. 1994. Evolution as Fact and Theory. In Gould's *Hen's Teeth and Horse's Toes*. New York: Norton.

Haught, John F. 2000. *God After Darwin: A Theology of Evolution*. Boulder, CO: Westview Press.

Haught, John F. 2001. *Responses to 101 Questions on God and Evolution*. Mahwah, NJ: Paulist Press.

Haught, John F. 2003. *Deeper Than Darwin*. Boulder, CO: Westview Press.

Isaac, Mark. 2003. *An Index to Creationist Claims*. Talk.Origins, http://www.talkorigins.org/indexcc/index.html.

Johanson, Donald, and Blake Edgar. 1996. *From Lucy to Language*. New York: Simon & Schuster.

Kitcher, Philip. 1982. *Abusing Science: The Case Against Creationism*. Cambridge, MA: MIT Press.

Larson, Edward J. 1997. *Summer for the Gods: The Scopes Trial and America's Continuing Debate over Science and Religion*. New York: Basic Books.

Larson, Edward J. 2003. *Trial and Error: The American Controversy over Creation and Evolution*, rev. ed. New York: Oxford University Press.

Lubenow, Marvin L. 1992. *Bones of Contention*. Grand Rapids, MI: Baker Book House.

Marsden, George M. 1991. *Understanding Fundamentalism and Evangelicalism*. Grand Rapids, MI: Eerdmans.

Matsumura, Molleen, ed. 1995. *Voices for Evolution*, 2nd ed. Berkeley CA: National Center for Science Education.

McGrath, Alister E. 1999. *Science & Religion: An Introduction*. Oxford: Blackwell.

Miller, Jonathan, and Borin Van Loon. 1982. *Darwin for Beginners*. New York: Pantheon Books.

Miller, Keith, ed. 2003. *Perspectives on an Evolving Creation*. Grand Rapids, MI: Eerdmans.

Miller, Kenneth R. 1994. Life's Grand Design. *Technology Review* 97 (2): 24–32.

Miller, Kenneth R. 1999. *Finding Darwin's God*. New York: HarperCollins.

Moore, John A. 1999. *Science as a Way of Knowing*. Cambridge, MA: Harvard University Press.

Moore, Randy. 2001. *Evolution in the Courtroom*. Santa Barbara, CA: ABC–Clio.

Morris, Henry M., ed. 1974. *Scientific Creationism*. San Diego: Creation-Life Publishers.

National Academy of Sciences. 1998. *Teaching About Evolution and the Nature of Science*. Washington, DC: National Academy Press.

National Academy of Sciences. 1999. *Science and Creationism: A View from the National Academy of Sciences*, 2nd ed. Washington, DC: National Academy Press.

Numbers, Ronald. 1992. *The Creationists*. New York: Knopf.

Numbers, Ronald L. 1998. *Darwinism Comes to America*. Cambridge, MA: Harvard University Press.

Pennock, Robert T. 1999. *Tower of Babel: The Evidence Against the New Creationism*. Cambridge, MA: Bradford Book/MIT Press.

Pennock, Robert T., ed. 2001. *Intelligent Design Creationism and Its Critics*. Cambridge, MA: MIT Press.

Plait, Philip. 2002. *Bad Astronomy*. New York: Wiley.

Pratt, H. Douglas. 2004. *The Hawaiian Honeycreepers*. Oxford: Oxford University Press.

Ridley, Mark, ed. 1996. *The Darwin Reader*, 2nd ed. New York: Norton.

Roberts, Jon H. 1988. *Darwinism and the Divine in America: Protestant Intellectuals and Organic Evolution, 1859–1900*. Madison: University of Wisconsin Press.

Ross, Hugh. 1994. *Creation and Time*. Colorado Springs, CO: NavPress.

Ruse, Michael. 1999. *The Darwinian Revolution: Science Red in Tooth and Claw*, 2nd ed. Chicago: University of Chicago Press.

Ruse, Michael. 2001. *Can a Darwinian Be a Christian?: The Relationship Between Science and Religion*. Cambridge: Cambridge University Press.

Ruse, Michael. 2003. *Darwin and Design: Does Evolution Have a Purpose?* Cambridge, MA: Harvard University Press.

Ruse, Michael, ed. 1996. *But Is It Science?* Amherst, NY: Prometheus.

Russell, Robert John. 2000. *Theology and Science: Current Issues and Further Directions*. Center for Theology and the Natural Sciences, 2000. Accessed October 21 2003. Available from http://ctns.org/russell_article.html.

Scott, Eugenie C. 1997. Antievolutionism and Creationism in the United States. *Annual Review of Anthropology* 26: 263–289.

Scott, Eugenie C. 2000. *The Creation–Evolution Continuum*. http://www.ncseweb.org/resources/articles/4606_the_creationevolution_continu_12_7_2000.asp.

Sober, Elliott. 2000. *Philosophy of Biology*, 2nd ed. Boulder, CO: Westview Press.

Stenger, Victor J. 2003. *Has Science Found God?* Amherst, NY: Prometheus Books.

Strahler, Arthur N. 1999. *Science and Earth History*. Amherst, NY: Prometheus Books.

Tudge, Colin. 2000. *The Variety of Life*. Oxford: Oxford University Press.

Van Till, Howard J. 2002. *E. Coli at the No Free Lunchroom: Bacteria Flagella and Dembski's Case for Intelligent Design*. Accessed November 30, 2002. Available from http://www.counterbalance.net/id-hvt/index-frame.html.

Weiner, Jonathan. 1994. *The Beak of the Finch: A Story of Evolution in Our Time*. New York: Knopf.

Wells, Jonathan. 2000. *Icons of Evolution: Science or Myth?* Washington, DC: Regnery.

Wexler, Jay. 1997. Of Pandas, People, and the First Amendment. *Stanford Law Review* 49: 439–470.

Wills, Christopher, and Jeffrey Bada. 2000. *The Spark of Life*. New York: Oxford University Press.

Witham, Larry. 2002. *Where Darwin Meets the Bible: Creationists and Evolutionists in America.* New York: Oxford University Press.

Zimmer, Carl. 2001. *Evolution: The Triumph of an Idea.* New York: HarperCollins.

Web Sites Cited

Access Research Network. http://www.arn.org.

American Experience. *Monkey Trial.* http://www.pbs.org/wgbh/amex/monkeytrial.

American Museum of Natural History, Paleontology Division. 2003. http://paleo.amnh.org/fossil/FRC.frontdoor.

Answers in Genesis. http://www.answersingenesis.org.

Behe's Empty Box. http://www.world-of-dawkins.com/Catalano/box/behe.htm.

Boston Review. 1996–2003. http://www.bostonreview.net/evolution.html.

British Broadcasting Corporation. *Darwin— the Man and the Legacy.* http://www.bbc.co.uk/education/darwin/leghist/index.htm.

British Library. *The Writings of Charles Darwin on the Web.* http://pages.britishlibrary.net/charles.darwin/.

Center for Renewel of Science and Culture. http://www.discovery.org/csc/TopQuestions/wedgeresp.pdf.

Center for Science and Culture. www.discovery.org/crsc.

Center for Theology and the Natural Sciences. http://ctns.org.

Counterbalance. http://counterbalance.org/.

Creation Research Society. http://www.creationresearch.org/.

Dialogue on Science, Ethics, and Religion. http://www.aaas.org/spp/dser/evolution/.

Evolution and the Nature of Science Institutes. http://www.indiana.edu/~ensiweb/home.html.

Foundation for Thought and Ethics. http://www.fteonline.com/.

Institute for Creation Research. http://www.icr.org.

Institute of Human Origins. 2003. *Becoming Human.* http://www.becominghuman.org.

Isaak, Mark. 2003. *An Index to Creationist Claims.* Talkorigins, http://www.talkorigins. org/indexcc/index.html.

Linder, Douglas. *Tennessee vs. John Scopes: The "Monkey Trial."* http://www.law.umkc.edu/faculty/projects/ftrials/scopes/scopes.htm.

Maddison, David R. 2003. *The Tree of Life Web Project.* http://tolweb.org/tree/phylogeny.html.

Metanexus Institute on Science and Religion. http://www.metanexus.org/.

National Academy Press. *Special Resources on Evolution.* http://lab.nap.edu/nap-cgi/discover.cgi?term=evolution&restric=NAP&ref=NAP.

National Center for Science Education. http://www.ncseweb.org.

Reasons to Believe. http://www.reasons.org.

Talk.Design. http://www.talkdesign.org.

Talk.Origins Archive. http://www.talkorigins.org.

University of California Museum of Paleontology. http://www.ucmp.berkeley.edu.

WGBH and Clear Blue Sky. *Evolution.* http://www.pbs.org/wgbh/evolution/.

Name Index

Subject Index

About the Author

EUGENIE C. SCOTT is Executive Director of the National Center for Science Education, the leading advocacy group for the teaching of evolution in the United States. She has written extensively on the evolution-creationism controversy in scholarly and popular venues, and she has won numerous awards for her work from scholarly organizations, including the Public Service Award from the National Science Board. She is a recent past-president of the American Association of Physical Anthropologists.